Nanotechnology for the Delivery of Therapeutic Nucleic Acids

Pan Stanford Series on Biomedical Nanotechnology

Series Editors

Vladimir Torchilin and Monsoor Amiji

Titles in the Series

Vol. 1

Handbook of Materials for Nanomedicine

Vladimir Torchilin and Monsoor Amiji, eds. 2010

978-981-4267-55-7 (Hardcover)
978-981-4267-58-8 (eBook)

Vol. 2

Nanoimaging

Beth A. Goins and William T. Phillips, eds. 2011

978-981-4267-09-0 (Hardcover)
978-981-4267-91-5 (eBook)

Vol. 3

Biomedical Nanosensors

Joseph Irudayraj, ed.
2013

978-981-4303-03-3 (Hardcover)
978-981-4303-04-0 (eBook)

Vol. 4

Nanotechnology for the Delivery of Therapeutic Nucleic Acids

Dan Peer, ed.
2013

978-981-4411-04-2 (Hardcover)
978-981-4411-05-9 (eBook)

Vol. 5

Inorganic Nanomedicine

Bhupinder Singh Sekhon, ed.
2014

Vol. 6

Nanotechnology for Cancer

Julia Ljubimova, ed.
2014

Vol. 7

Nanotechnology for Delivery of DNA and Related Materials

Bengt Fadeel, ed.
2015

Vol. 8

Translation Industrial Nanotechnology

Thomas Redelmeier, ed.
2015

Nanotechnology for the Delivery of Therapeutic Nucleic Acids

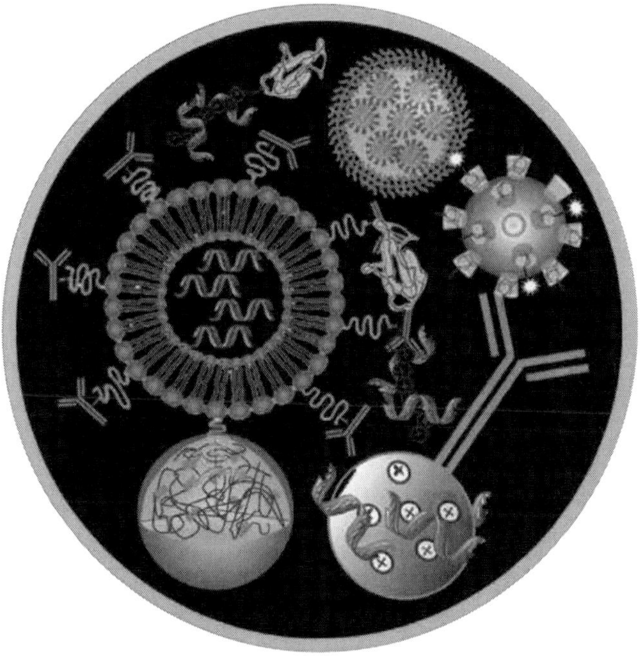

Dan Peer
Editor

Pan Stanford Publishing

Published by

Pan Stanford Publishing Pte. Ltd.
Penthouse Level, Suntec Tower 3
8 Temasek Boulevard
Singapore 038988

Email: editorial@panstanford.com
Web: www.panstanford.com

British Library Cataloguing-in-Publication Data
A catalogue record for this book is available from the British Library.

Nanotechnology for the Delivery of Therapeutic Nucleic Acids

ISBN 978-981-4411-04-2 (Hardcover)
ISBN 978-981-4411-05-9 (eBook)

Printed in the USA

Contents

Preface

More than three decades ago, Paul Zamecnik and his colleagues suggested that nucleic acids (NA) could be used to block gene function by virtue of Watson–Crick base pairing. Since the discovery of RNAi in 1998 by Andrew Fire and Craig Melo and soon after the discovery that RNAi is found in mammals in 2001 by Thomas Tuschl's group, synthetic small RNAs were shown to treat disease in mice. Small RNAs were quickly proclaimed as the "next new class of drugs." Eagerness sprinted high because of the potential of these molecules to knock down any gene of interest to treat almost any disease by targeting otherwise "undruggable" targets such as molecules without ligand-binding domains or enzymatic function. Despite the promise, developing any NA as therapeutics has proven challenging. Like most drug development, there is no quick fix. Although many of the hurdles to developing NA-based drugs have been easily addressed, the main obstacle is figuring out how to deliver these molecules into cells in a therapeutically acceptable way. Small RNAs being considered therapeutic drugs include not only siRNAs designed to knock down one gene at a time but also mimics of endogenous microRNAs to suppress the expression of many genes, but with less efficient suppression of each one. The delivery hurdle that needs to be solved to administer siRNAs and imperfectly paired microRNA mimics is essentially the same (although antagonizing endogenous micro RNAs using single-stranded antisense oligonucleotides may be somewhat easier). When injected intravenously, NA are rapidly cleared by renal filtration and are susceptible to degradation by extracellular RNases or DNases. The NA half-life can be increased—even to days—by chemical modifications to eliminate susceptibility to endogenous exonucleases and endonucleases and by incorporating the NA into a larger moiety, above the molecular weight cutoff for kidney filtration. However, entering the cell is the biggest obstacle. Because of their large molecular weight and net negative charge, naked NA do not cross the plasma membrane. Although cells can endocytose many types of modified NA or NA-containing particles, another important bottleneck is getting these molecules efficiently out of the endosome into the cytosol where the RNAi machinery resides or into the nucleus for DNA to work.

NA therapeutics has been extensively studied both in the academia and in the pharmaceutical industry and is still considered the promise for new therapeutic modalities, especially in personalized medicine. The only hurdle that limits the translation of NA therapeutics from an academic idea to new therapeutic modality is the lack of efficient and safe delivery strategies. In this book, written by world experts in the field of nanotechnology for NA delivery, we bring together the state of the art in delivery strategies using lipids, polymers, chemical conjugates, NA aptamers, and proteins with strong emphasis on issues and aspects that are of essence to the pharmaceutical industry working in this area such as stability, general toxicity, immune-toxicity, pharmacokinetics and naturally efficacy and validation of new drug targets *in vivo* using unique approaches based on exquisite nanotechnology strategies.

The work by Prof. Gerardo Byk and colleagues (Chapter 1) provides a tutorial overview of lipoplex and polyplex from a chemical standpoint. Discussions about lipopolyamines, lipopoly-aminoguanidines, and reduction-sensitive lipopolyamine and dendrimers provide new insights into chemical modifications toward non-electrostatic DNA complexing agents.

The work by Prof. Avi Domb and colleagues (Chapter 2) provides an excellent overview on the major cationic polymers used for the delivery of nucleotides, among them polyethylenimine, poly (L-lysine), cationic polysaccharides (such as chitosan, cyclodextrins, and dextran-spermine), dendrimers, cationic polyesters, poly(amino ester)s, and poly(amido amine)s. Factors influencing cationic polymer-mediated nucleotide delivery are also discussed. In addition, several biomedical applications are discussed, such as siRNA delivery, DNA vaccination, lung and liver delivery, brain delivery, and tumor delivery.

Chapter 3, authored by Prof. Leaf Huang and colleagues, provides an introduction to the challenges in nanocarriers systems for NA delivery. It details two strategies of membrane/core NPs based on lipids, the LPD, and the LCD and discusses several applications in siRNA delivery using these strategies.

Another interesting strategy is the delivery of single siRNA molecules by peptide transduction domains as described by Steve Dowdy and colleagues in Chapter 4. Additional RNA binding proteins are also detailed.

Chapter 5, written by Prof. John Rossi and colleagues, reviews the current advances of cell-specific aptamers in cell recognition and targeted delivery, with a particular focus on the development of the aptamer-functionalized siRNA or nanocarrier for targeted gene silencing.

Prof. Ken Howard details in Chapter 6 bioresponsive nanoparticles based on copolypeptides and hyperbranched polymers for controlling the intracellular spatial and temporal effects of synthetic microRNA and siRNA.

In Chapter 7, Prof. Robert Langer and Prof. Michael Goldberg describe the synthesis, screening, formulation, evolution, and application of "lipidoids," a novel class of lipid-like molecules that highlights the utility of combinatorial approaches for the production of effective siRNA delivery vehicles.

My personal contribution to this book is Chapter 8, in which I detail the use of integrin targeted and stabilized lipid-based nanoparticles for the manipulation of leukocytes' function using RNAi.

Finally, Prof. Nigel McMillan and his colleagues outline efforts to improve not only delivery but also RNAi efficacy in the vaginal mucosa as a means to treat genital infections, particularly virally driven cervical cancer, using various strategies.

Clear, easy to understand, and focused on key issues for future research and development, this book provide new insights into the dynamic field of NA delivery using nanotechnology.

I am grateful to all the authors who contributed to this book, among them Prof. Byk, from Bar-Ilan University, Prof. Domb from the Hebrew University in Jerusalem, Prof. Huang from the University of North Carolina at Chapel Hill, Prof. Dowdy from the University of California San Diego, Prof. Rossi from the City of Hope in California, Prof. Howard from the University of Aarhus, Prof. Langer from MIT, Prof. Goldberg from Harvard Medical School, and Prof. McMillan from the University of Queensland.

Special thanks to my wife, Shlomit, and my children, Dor, Barak, and Naama, for their unrestricted support.

This book is dedicated to the memory of my parents, Itta and Alexander Peer, who educated me to strive for knowledge and excellence.

Dan Peer
Tel Aviv, Winter 2012

Chapter 1

Lipoplexes and Polyplexes: From Gene Delivery to Gene Expression

Gerardo Byk,*[*] Mirit Cohen-Ohana, and Fiana Mirkin

Laboratory of Nano-Biotechnology, Department of Chemistry, Bar Ilan University, Ramat Gan 52900, Israel

[*]gerardo.byk@biu.ac.il

1.1 Introduction

The human genome is rapidly being decoded. A map of the human genome composed of about 30000 genes was presented some years ago [1, 2]. Together with these findings, an increasing number of genes have emerged that mediate both genetic and acquired diseases. These dramatic advances make gene therapy an important avenue for the treatment of genetic human diseases as well as major acquired killers like cancer and AIDS. Therefore, having access to reliable gene transfer techniques is of critical importance.

Unfortunately, even after about 25 years of research, gene transfer is still a rather inefficient process hampered by several limitations intrinsically paired to the specific gene delivery approaches. A breakthrough in gene therapy clinical trials will probably emerge from combination and application of substantially different gene delivery techniques based on viral and synthetic vectors.

Nanotechnology for the Delivery of Therapeutic Nucleic Acids
Edited by Dan Peer

ISBN 978-981-4411-04-2 (Hardcover), 978-981-4411-05-9 (eBook)
www.panstanford.com

The most efficient gene delivery systems for foreign gene transfer to eukaryotic cells are viruses. In viral vectors, the genetic material is part of the genome of a replicative–defective virus. Penetration, integration, and transcription in the host cell via the viral natural pathway lead to foreign gene expression. Although a variety of viral vectors were developed during the last 25 years [3, 4] and clinical trials using viral vectors have been accomplished [5], their use is hampered by limitations such as the size of the foreign gene to be transferred, the risk of intrinsic viral propagation and immunogenicity [6, 7].

Therefore, in order to circumvent the drawbacks associated with viral vectors, a variety of non-viral gene delivery systems have been developed.

Synthetic gene delivery systems will be, in principle, devoid of propagation risks, will not induce immune responses, and will not be limited in the size of the foreign gene to be expressed. However, their capacity to transfect cells by a process that includes tissue-specific targeting, cell penetration, transgene integration, and/or transcription remains to be improved. Synthetic gene delivery systems are more a complementary option to virus-mediated gene therapy of human diseases rather than a separate approach. For example, after a first treatment with a viral vector that might palliate but not cure a given disease, an immune response might be induced. The synthetic DNA delivery agent can be useful for a complementary second-round treatment, instead of using a second dose of the viral vector, whose efficacy might be compromised by the immune response switched-on during the first treatment. Synthetic DNA delivery agents are especially recommended for delivering DNA to tissues resistant to viral gene transfer [8]. Moreover, various synergistic systems have been described, exploiting the advantages of both viral and non-viral gene delivery systems, such as adeno-lipofection or retro-lipofection in which infection with adeno virus [9–12] or retrovirus [13–15] is performed in the presence of cationic lipid vectors in tissues, where classical viral transfection is poor, resulting in improved transgene expression. Another synergy between different gene transfer methodologies is exemplified by the plasmovirus [16]. Plasmoviruses are plasmids capable of expressing all the viral genes required for generating infectious particles and packaging a defective genome containing a transgene. Plasmids transfected using cationic lipids transform the transduced cells into packaging cells that release infectious replication-defective retrovirus vectors (RV) containing a transgene, which will infect nearby neighbor cells. Such a vector can efficiently "propagate" the transgene after transfection with synthetic vectors. This system is especially suited for suicide gene strategies [17]. Finally, the advent of small interfering RNA (siRNA), a class of double-stranded RNA nucleotides composed of 20–25 bases, with a variety of roles in biology, is involved in the RNA interference (RNAi) pathway, where it interferes with the expression of a specific gene [18]. Therefore, siRNA can be used for treating genetic diseases mediated by a pathogenic protein that, once silenced by an appropriate siRNA, can be palliated. An advantage of siRNA is that its low molecular weight allows a better bioavailability as compared to plasmid DNA and its complexion with synthetic gene delivery carriers improves its bio-stability.

In general, complexes formed between the positively charged synthetic vectors and the negatively charged genetic material (either plasmid DNA or siRNA) are mostly characterized as being nanometric multi-lamellar bilayer particles (see Fig. 1.1).

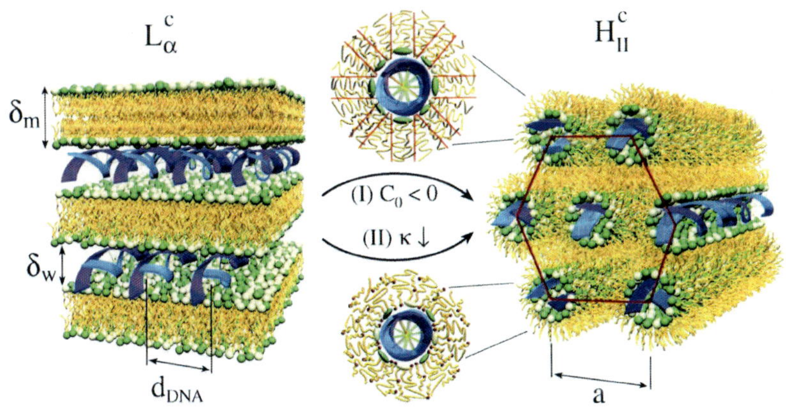

Figure 1.1 Schematic of two distinct pathways from the lamellar L_α^C phase to the columnar inverted hexagonal H_{II}^C phase of cationic lipid-DNA complexes. Along pathway I, the natural curvature C_0 5 $1/R_0$ of the cationic lipid monolayer is driven negative by the addition of the helper-lipid DOPE. This is shown schematically (center top); the cationic lipid DOTAP is cylindrically shaped whereas DOPE is cone-like, leading to the negative curvature. Along pathway II, the $L_\alpha^C \rightarrow H_{II}^C$ transition is induced by the addition of helper-lipids consisting of mixtures of DOPC and the co-surfactant hexanol, which reduces the membrane bending rigidity. This figure was reproduced with permission [19].

The +/– cationic lipid/DNA charge ratio predetermines the size and shape of the formed complexes. At low +/– charge ratio, DNA molecules are partially "decorated" by the cationic lipid, thus a strong propensity to aggregation is observed in the complexes that result in micrometric particles with low transfection efficiency. On the other hand, at high +/– charge ratio, DNA is highly packed in multi-lamellar bilayer particles of about 90–300 nm as shown in Fig. 1.1. These types of particles are the most efficient for gene delivery to cells.

For more than 10 years, the design, synthesis, and characterization of novel cationic lipids and polymers for increased gene delivery and transgene expression have been our central preoccupation. Most of the topics summarized here deal with lipopolyamines/cationic lipids or systems containing them (lipoplexes) and some polymers (polyplexes) or dendrimers (dendriplexes).

1.2 Lipopolyamines

Cationic lipids are composed of three elements: from one side a cationic entity, for example amines, quaternary ammonium salts, etc., from the other side a lipid entity usually a double lipid chain, cholesterol, etc., and, in between, a linker entity that brings together the other elements. In our works, we have implemented a step-by-step strategy by introduction of systematic modifications into the cationic lipid backbone and the introduction of a "side chain entity" on the linker (see Fig. 1.2).

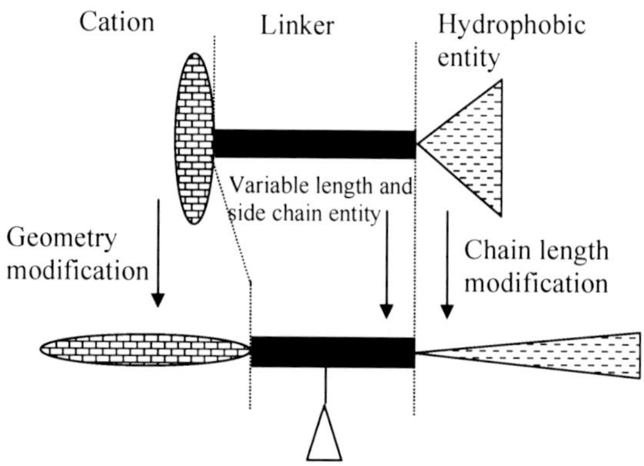

Figure 1.2 Structure–activity relationship studies of lipopolyamines.

Our first contribution to this field was a novel family of geometrically differing lipopolyamines [20–22]. We have exploited, for the first time, solid-phase combinatorial chemistry to obtain a variety of geometrically varied mono-functionalized polyamines as building blocks for generating the structurally different lipopolyamines. The use of a solid phase support for the synthesis of geometrically varied polyamines significantly simplifies the number of steps in the synthesis [23–25]. Thus, taking advantage of a dilution effect obtained by linking a bromo-acetyl group to a solid phase support, a variety of mono-functionalized protected polyamines were rapidly obtained in a single step without isolating any intermediate. After cleavage from the solid support, the free carboxylic function of the building blocks was reacted with the appropriate linkers and lipids to obtain the desired products (see Fig. 1.3).

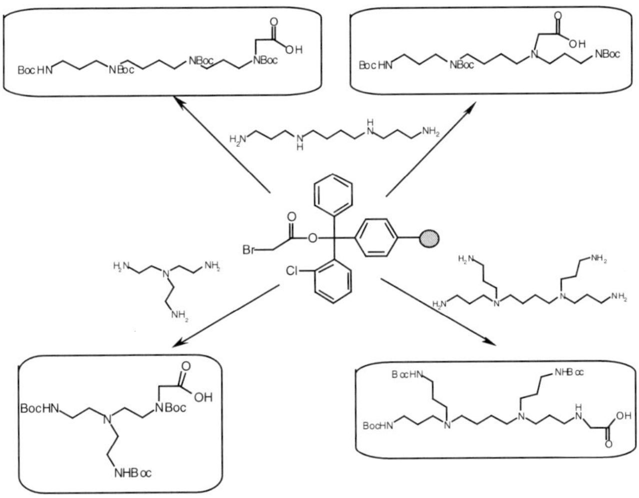

Figure 1.3 Synthesis of geometrically varied mono-functionalized polyamine building blocks for the synthesis of geometrically varied lipopolyamines.

The polyamine, the spacer, and the lipid moiety of the lipopolyamines were systematically modified, and the impact of these modifications on gene delivery was demonstrated *in vitro* on several cell lines and *in vivo* for selected products [20]. The first family consisted of geometrically differing polyamines described in Fig. 1.4.

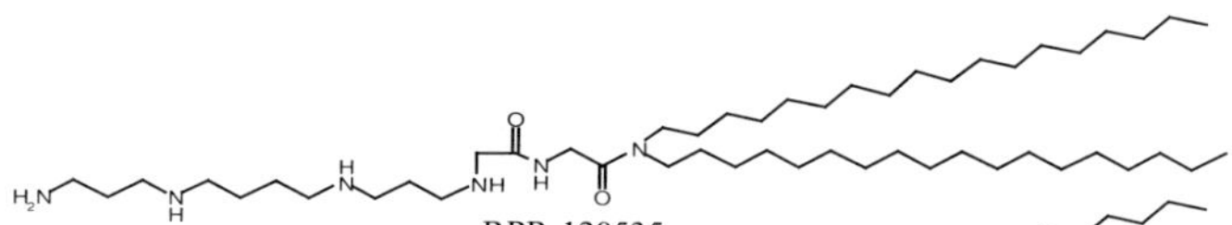

RPR-120535

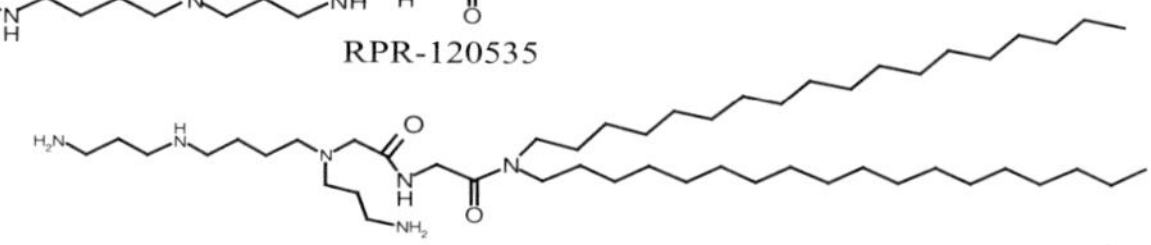

RPR-126096

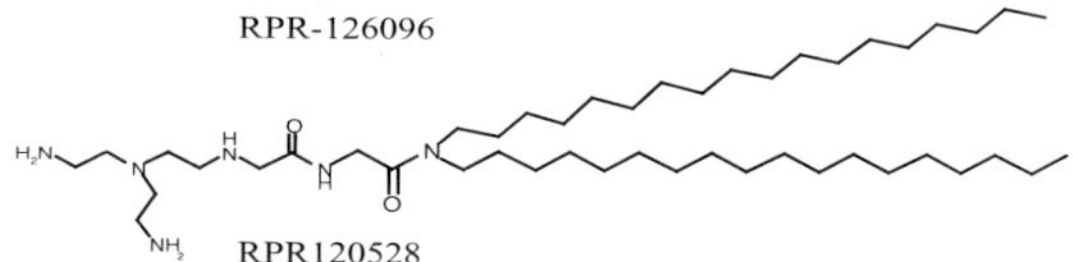

RPR120528

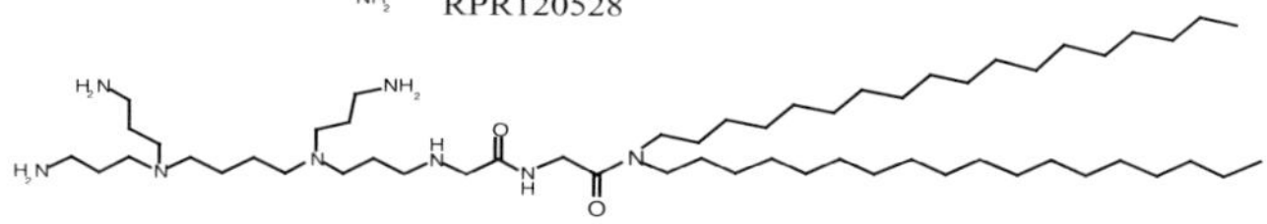

RPR120525

Figure 1.4 Geometrically varied lipopolyamines for structure–activity relationship studies.

The obtained lipopolyamines were assayed *in vitro* on HeLa cells. Results obtained in Table 1.1 show that although all the lipopolyamines displayed high level of transfection, the "linear" RPR-120535 showed the highest level of luciferase expression, followed by the "branched" RPR-120525 and "« T » shaped" RPR-126096, and finally the "globular" lipopolyamine RPR-120528.

Table 1.1 Impact of polyamine geometry on expression of luciferase on HeLa cell line in the presence or absence of serum. Results are in RLU/10s./2 × 10^3 treated cells

N° RPR	Geometry	HeLa
120535	Linear	2.4×10^6
126096	« T » shape	5.5×10^5
120528	Globular	1.3×10^5
120525	Branched	5.8×10^5

In agreement with previous works [26], the geometry of the cationic entity affects the transfection efficiency of the self-assemblies with DNA.

The introduction of a variety of linkers between the lipid and the polyamine did not affect substantially the transfection efficiency as compared to the best head of series RPR-120535. On the other hand, when the C-18 lipid chain in RPR-120535 was replaced by shorter chains of C-14, C-13, or C-12, a significant lose in transfection efficiency of one, two, or three orders was observed concomitantly [20].

Another novelty of this approach was the introduction of a fourth, new element in the lipopolyamine backbone called "side chain entity." This side chain could play various roles such as targeting, labeling, or stabilizing lipid/DNA complexes. This approach was demonstrated by the synthesis of a series of lipopolyamines bearing a variety of side chains suitable for targeting (see Fig. 1.5) (biotinyl [RPR-122761], arachidonyl [RPR-130605], and glycosyl [RPR-130596]), for labelling (rhodaminyl [RPR121653]), or for physico-chemical stabilization of complexes with DNA (guanidyl [RPR120531 and RPR121650]). Thus, it was demonstrated that the introduction of a linker bearing a side chain entity is allowed for transfection *in vitro* and *in vivo*. Additionally, the introduction of a molecular probe, such as rhodaminyl, allowed to investigate the intracellular fate of DNA/cationic lipid complexes [27].

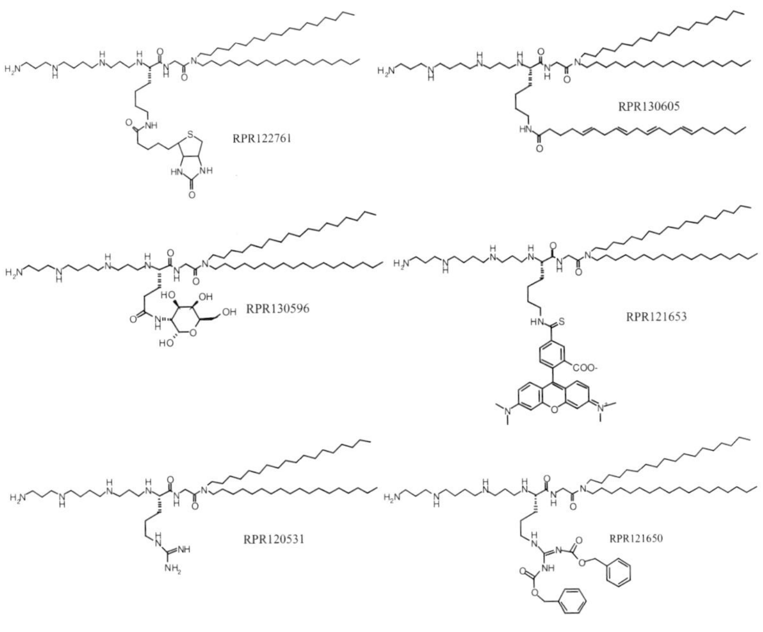

Figure 1.5 Introduction of side chain entities into lipopolyamines.

Extensive physicochemical studies on RPR120535, a leading product of the series, demonstrated the presence of characteristic multi-lamellar bilayers, particles of 200 nm, with a surprising periodicity of 80 A [28]. Unlike quaternary ammonium lipids, these multi-lamellar particles are formed in the presence or absence of any helper or additive lipid, such as DOPE. In an extensive study, the structural polymorphism of DNA/RPR120535 complexes has been studied by X-ray diffraction and cryo-electron microscopy (see Fig. 1.6). Monovalent salts and temperature effects have been

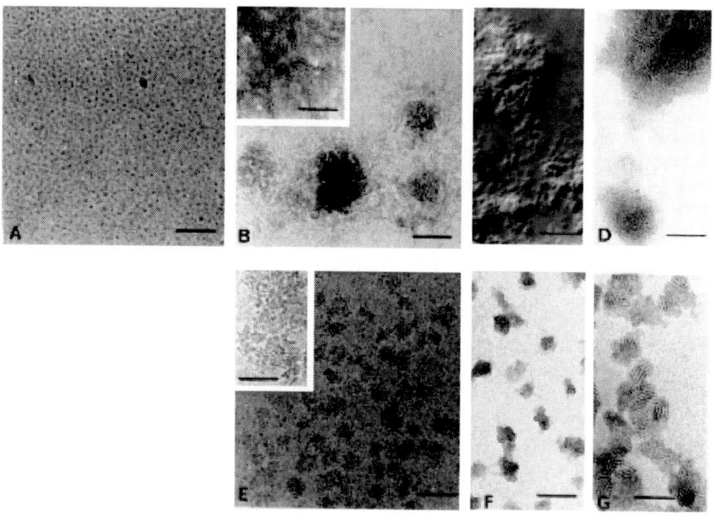

Figure 1.6 RPR120535 and RPR120535/DNA complexes from zones A, B, and C, visualized by light microscopy (TEM). (A) Cryo-TEM micrographs of an aqueous solution of RPR120535 alone. (B) Electron micrograph of uranyl acetate-stained complexes from zone A (Charge ratio +/–: 0.3). Inset shows the same complexes at higher magnification. (C) Complexes from zone B (Charge ratio +/–: 1.65) observed by light microscopy. (D) Same complexes as (C) stained with uranyl acetate. (E) Cryo-TEM micrograph of RPR120535/DNA complexes from zone C (Charge ratio +/–: 6). Inset shows the visualization by cryo-TEM of the ordered domains in these complexes. (F) Electron micrograph of uranyl acetate stained complexes from zone C. (G) Cryo-phosphotungstate-TEM micrograph of the same complexes. In this micrograph, the complexes seem to have aggregated, because the thickest part of the vitrified film allows them to be superimposed. The scale bar represents 100 nm in A, B inset, and D–G; 500 nm in B; and 10 micron in C. Reproduced with permission [28].

In further works [12], we have used compound RPR-120535 *in vivo* for correcting anemia in a uremic-mice model by gene delivery and expression of erythropoietin (EPO) to peritoneal mesothelial cells that, upon gene delivery, secreted EPO to blood.

Finally, MCO-I-68, a lipopolyamine with a gadolinium complexing arm on its side chain (see Fig. 1.7), has been synthesized as probe for *in vivo* distribution studies of DNA/lipopolyamine complexes using MRI technics [31].

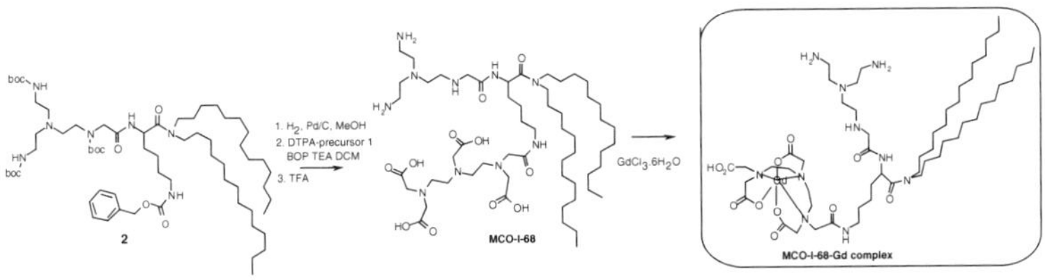

Figure 1.7 Synthesis of MCO-I-68 and its complex with gadolinium for MRI.

Results in panels A and B in Fig. 1.8 show the clear contrast obtained even after 48 h after administration of two different formulations of MCO-I-68.

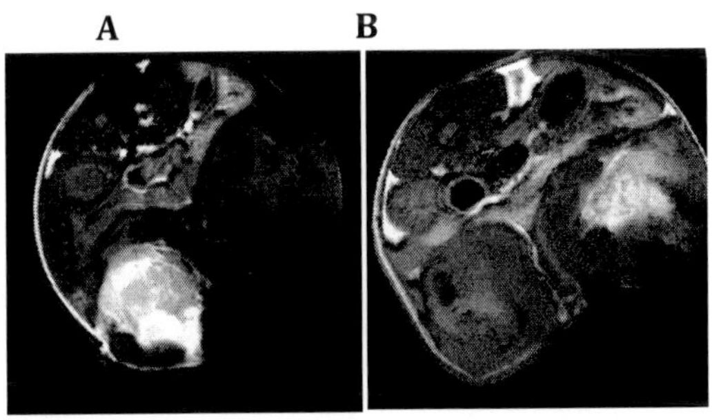

Figure 1.8 MRI Images of mice. A: 48 h after injection of the left side tumor with MCO-I-68-Gd/DOPE/ DNA complexes. Right side tumor is the non-injected tumor control. B: 48 h after injection of the right side tumor with MCO-I-68-Gd/DNA lipoplex.

1.3 Lipopolyamine Co-formulation with DNA Complexing Peptides

In the second approach, we have devised a system in which DNA is associated with short peptides derived from human histone or protamine, prior to the addition of a cationic lipid or polymer. The gold standard lipopolyamine RPR-120535 from the first approach has been co-formulated with a histone H1-derived peptide. The peptide-DNA-lipid particles displayed enhanced *in vitro* transfection efficiency over that observed with classical DNA/lipid complexes and in particular conferred transfection capacity in the presence of serum. This serum resistance is cell-type independent and was observed with a variety of lipopolyamines. Pre-compacting DNA with a histone H1-derived peptide enhances gene transfer in an *in vivo* model of Lewis lung carcinoma [32].

1.4 Lipopolyaminoguanidines

Further, we have introduced amidinium moieties into cationic lipids by introducing amidino-moieties through the linker or taking advantage of a similar combinatorial approach to that applied for obtaining polyamine-building blocks. In the first series of molecules, we introduced guanidinium groups in the position of the linker between the lipid and the polyamine [20–23, 25]. Product #8 (Fig. 1.9) displayed enhanced transfection activity compared to Lipofectamine on NIH3T3, Rabbit SMC, 3LL Lewis lung, and CaCo2 colon carcinoma cell lines. Product RPR120531 displays high level of transgene expression that is not inhibited by the presence of serum [27].

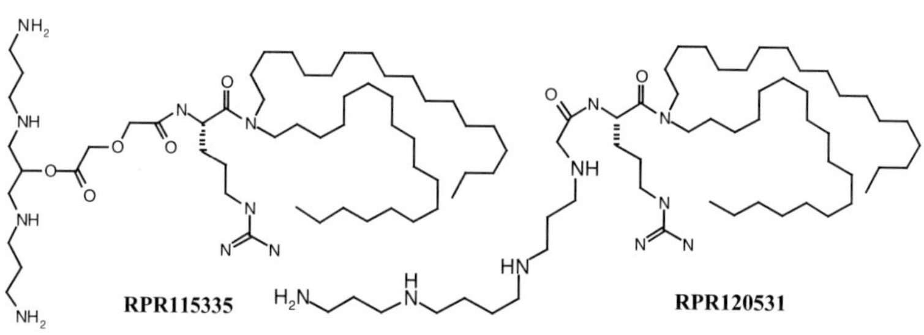

Figure 1.9 Lipopolyamines bearing a guanidinium function as side chain.

We then extended the family of poly(guanidinium)amine-lipids exploiting combinatorial chemistry technology. A second-generation library of mono-functionalized poly(guanidinium)amines were synthesized and introduced into cationic lipids applying the concept of "Libraries from Libraries," (Fig. 1.10) [21].

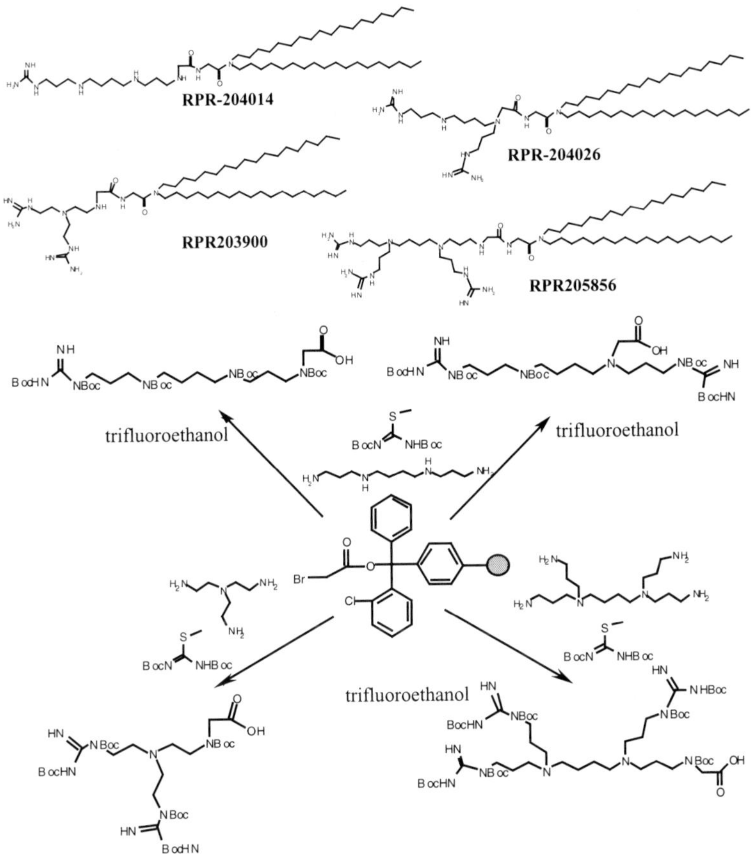

Figure 1.10 Synthesis of geometrically varied mono-functionalized poly(amino)-guanidine building blocks on solid phase (Libraries from libraries) and geometrically varied lipo-poly-guanidinium-amines containing the synthetic building blocks.

Finally, we have developed a family of cyclic polyaminoguanidines. Again a combinatorial approach applying the concept of libraries from libraries generated on solid support of a variety of building blocks that were introduced into lipids (Fig. 1.11). Some of the products have been found to be useful for intramuscular gene delivery and to be more efficient than naked DNA [33].

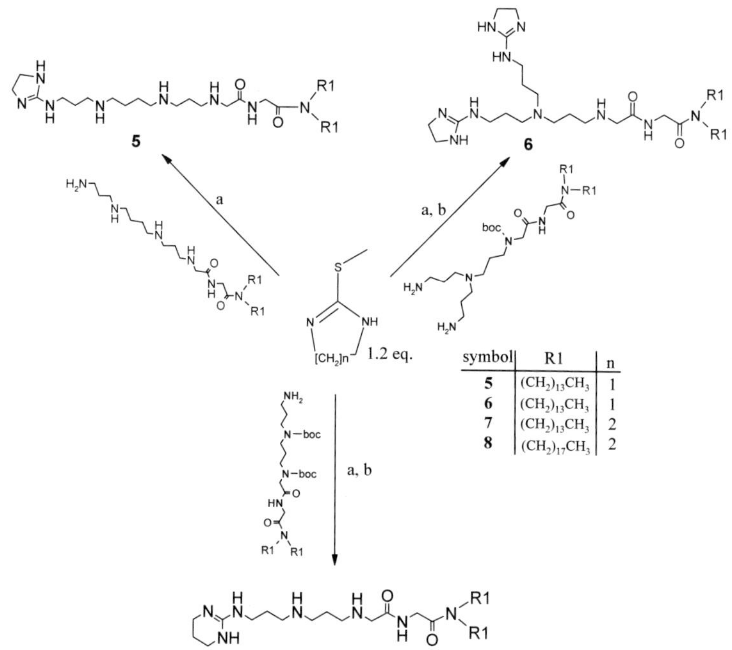

symbol	R1	n
5	$(CH_2)_{13}CH_3$	1
6	$(CH_2)_{13}CH_3$	1
7	$(CH_2)_{13}CH_3$	2
8	$(CH_2)_{17}CH_3$	2

Figure 1.11 Synthesis of lipopolyamino-cycloguanidines. (a) dichloromethane, TEA (1,3 eq), 20°C, overnight, HPLC purification. (b) TFA/DCM (1:1) 1 h 20°C (see Ref. [33] for detailed experimental procedure).

1.4.1 Biodegradable Lipoplexes: Reduction-Sensitive Lipopolyamines

We have designed and synthesized original lipopolyamines for modulated release of DNA from cationic lipid/DNA complexes [22, 34–36]. Our rationale was that modulated degradation of the lipids during or after penetration into the cell could improve the trafficking of DNA to the nucleus, resulting in increased transgene expression. The new reduction-sensitive lipopolyamines (RSL) harbor a disulfide bridge within different positions in the backbone of the lipids as bio-sensitive function. A useful synthetic method was developed to obtain unsymmetrical disulfide bridged molecules with very good yields and reproducibility, starting from symmetrical disulfides and thiols [22]. The new lipopolyamines are good candidates as carriers of therapeutic genes for *in vivo* gene delivery. To optimize the transfection efficiency in these novel series, we have carried out structure–activity relationship studies by placing the disulfide bridge at different positions in the backbone of the cationic lipid and by systematic variation of the lipid chain length and nature. Results indicate that transfection levels can be modulated as a function of the location of the disulfide bridge in the molecule (see Figs. 1.12 and 1.14).

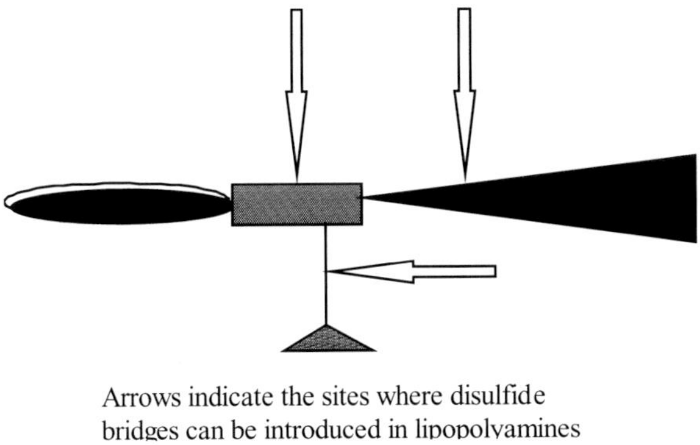

Figure 1.12 Structure–activity relationship studies on reduction sensitive lipopolyamines.

The synthesis is based on a convergent approach. In the first step, building blocks 1, 2, and 3 were prepared as described in Fig. 1.13.

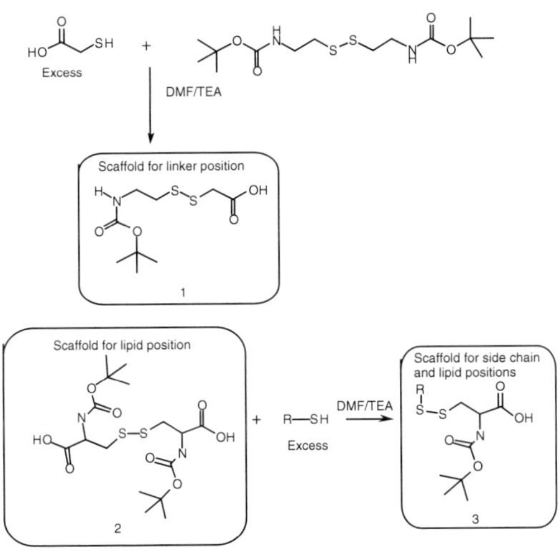

Figure 1.13 Synthesis of reduction-sensitive-lipopolyamine building blocks.

The building blocks were coupled to the corresponding lipids and functionalized polyamine-building blocks used for the synthesis of RPR120535 (Figs. 1.3 and 1.4), by successive peptide couplings and deprotection of Boc groups. The structures of the compounds are drawn in Fig. 1.14.

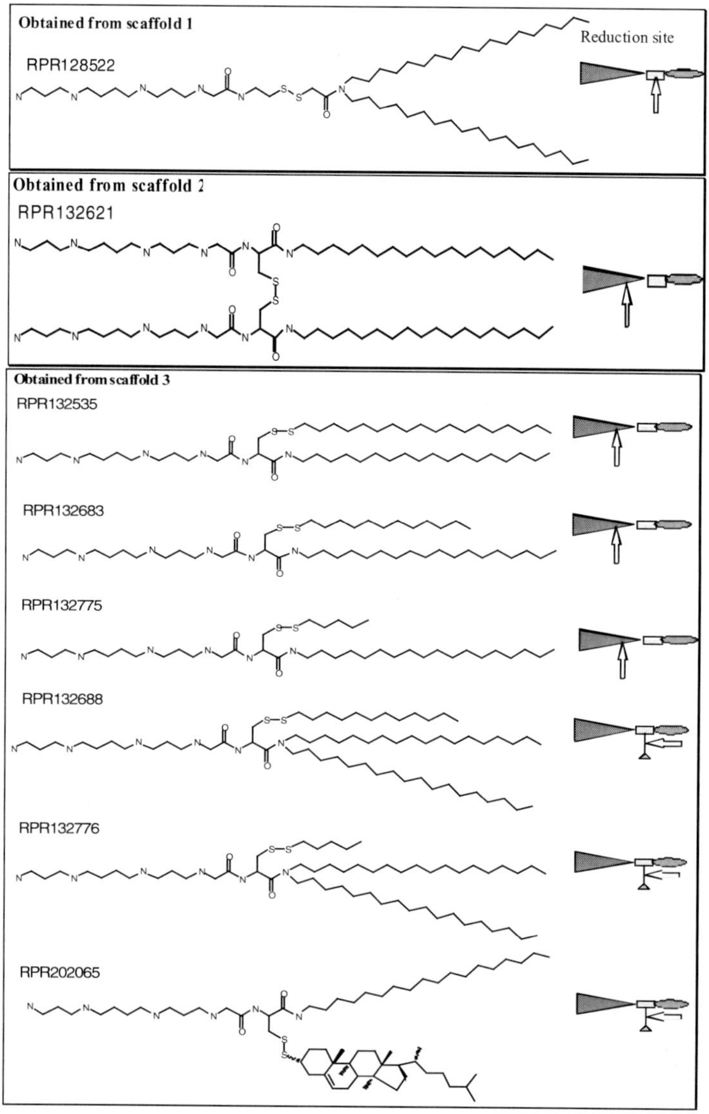

Figure 1.14 Structure of reduction-sensitive lipopolyamines.

The present series harbor disulfide bonds in every important position of the cationic lipid. Similarly to RPR120535, these cationic lipids are not formulated with DOPE or other additive for optimal transfection activity. Exhaustive structure–activity relationships studies were accomplished. From one side, results indicate that compound RPR-128522 with a disulfide bridge at the linker position is completely devoid of transfection activity, probably due to an early release of the DNA from complexes upon cell penetration. From the other side, RPR-132688 with a triple lipid chain, but only one of them cleavable by reductive conditions, is 10 times more active than RPR120535 (previously described as being 100–1000 times more efficient than first generation quaternary ammonium salt lipids such as DOTAP/DOPE depending on the cell line model). Physico-chemical characterization studies of the complexes [34–36] demonstrate that DNA release from complexes can be modulated as a function of the surrounding reducing conditions of the complexes and of the localization of the disulfide bridge within the lipopolyamine.

Overall we have shown that modulation of gene delivery and expression can be accomplished by proper introduction of *in situ*-reducible disulfide bridges at key positions in the backbone of the lipopolyamine. Our results suggest that RSL is a promising new approach for gene delivery.

1.4.2 Biodegradable Polyplexes: Reduction-Sensitive Dendrimers

Although non-viral DNA delivery agents are devoid of immunogenicity, they display high toxicity in many cases. This is especially significant in the case of polyplexes such as polylysine and cationic dendrimers, which were extensively used for non-viral gene delivery, and some of them were abandoned due to high toxicity. The integration of a biodegradable entity into non-viral gene delivery systems promises improved gene expression and decreased toxicity in the attended area as a result of the modulated release of DNA from complexes induced by the new biodegradable system [34]. We have designed and synthesized biodegradable polymers derived from polylysine, polyarginine, and polyhistidine of various lengths containing disulfide bridges along the backbone. Additionally, a biodegradable dendrimer was synthesized [37]. These polymers form complexes with DNA similarly to those obtained with non-degradable polycationic systems. After penetration into the cell the disulfide bridges are progressively degraded by the reducing conditions found at the different cytosolic compartments, inducing the release of DNA from complexes. The polymer is finally degraded to nontoxic monomers or peptides, and DNA is released for transfection.

Building block **SSa** was synthesized and introduced along the peptide chains of various polypeptides using conventional SPPS (see Fig. 1.15).

SSa = Fmoc-$NHCH_2CH_2$–S–S–$CH_2CH_2NHCOCH_2CH_2COOH$

Lys$_n$ $n = 1, 2, 3$ H–(K_{10}**SSa**)$_n$$K_{10}NH_2$

His$_n$ $n = 1, 2$ H–(H_{10}**SSa**)$_n$$H_{10}NH_2$

Figure 1.15 Reduction sensitive cationic polypeptides synthesized using building block **SSa.**

In another approach, we have developed a new method for solid-phase inter-site disulfide bond formation that resulted in dimers containing cysteine bridges. These dimers were generated after conventional SPPS of the polypeptides containing (S-trityl)cysteine at different positions. The cysteine residues were bridged on the solid support using iodine in methanol (see Fig. 1.16).

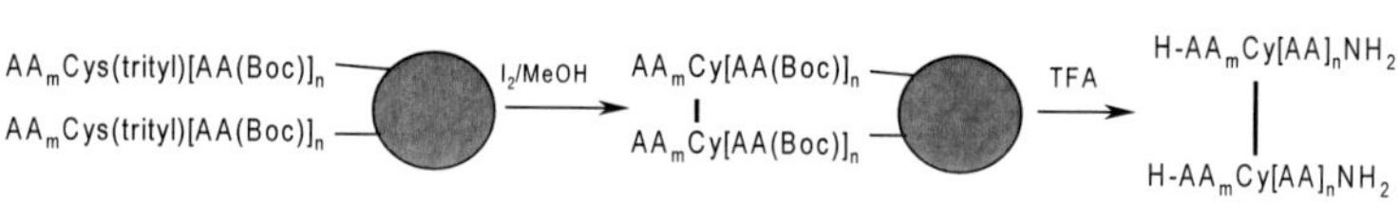

Figure 1.16 Reduction-sensitive cationic polypeptides obtained by solid-phase inter-site disulfide bridge formation: AA = K (**CyK**, $m = 0, n = 10$; **KcyK**, $m = 8, n = 2$); H (**CyH**, $m = 0, n = 10$; **HCyH**, $m = 8, n = 2$); R (**CyR**, $m = 0, n = 10$).

Additionally, we have combined the use of building block SSa and the inter-site disulfide bridge formation to obtain the long cationic polypeptide containing 40 histidines, FIM-2256: (H–H_{10}–**SSa–Cy**–$H_{10}NH_2$)$_2$.

Finally, we have synthesized the dendrimer FIM-2126 that contains a disulfide bridge (see Fig. 1.17).

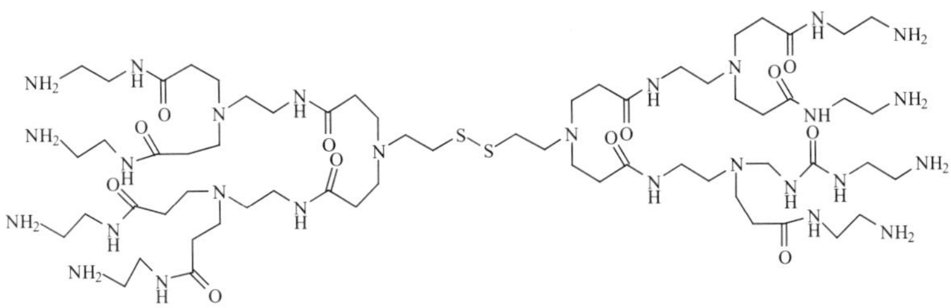

Figure 1.17 Structure of reduction-sensitive biodegradable dendrimer FIM-2126.

The products were mixed with DNA at different charge ratios and were run on gel electrophoresis in the presence or absence of dithiothreitol. Figure 1.18 shows that the products retain the DNA at positive charge ratios similarly to non-reducible polycations. On the other side, incubation of the complexes with DTT disrupts completely the complexes, and DNA runs freely independently of the charge ratios.

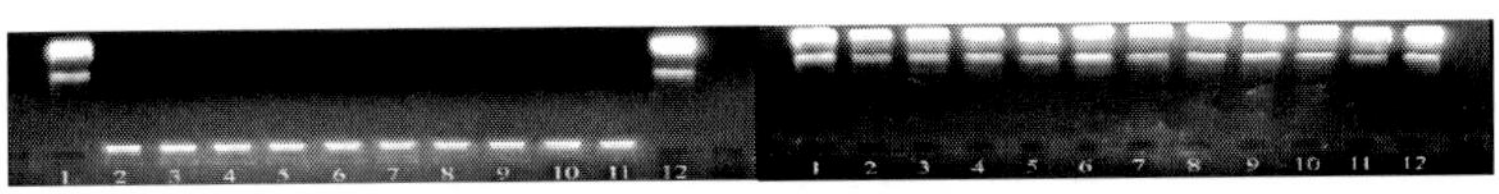

Figure 1.18 Gel electrophoresis obtained for DNA complexes of the polypeptides in absence (left panel) or presence of DTT (right panel). Lanes 1 and 12 stand for uncomplexed DNA.

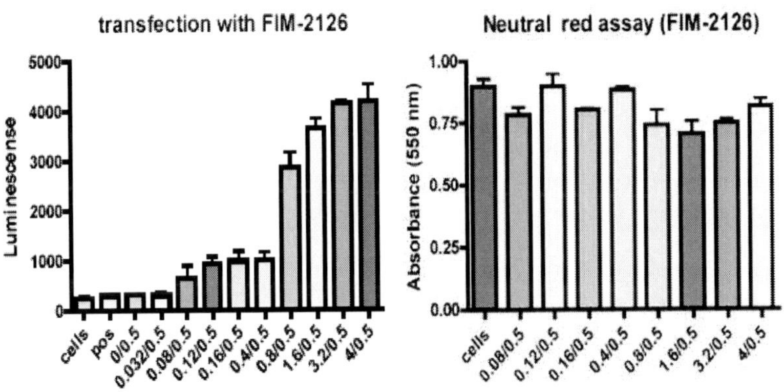

Figure 1.19 Left panel: Transfection levels in 3T3-BALB/c fibroblasts for dendrimer FIM-2126/SEAP-DNA complexes at different charge ratios (+/−). Right panel: XTT viability assay in 3T3-BALB/c fibroblasts of dendrimer FIM-2126 at different charge ratios (+/−).

These results proved *in vitro* that the reducible complexes form normal complexes with DNA but when treated with reducing conditions the complexes disrupt as expected. The complexes were tested for expression of SEAP after incubation with 3T3-BALB/c fibroblasts accompanied by cell viability assays using XTT kit as compared to the known standard RPR-120535 (see Fig. 1.19). All the polypeptides displayed low level of transfections (below 10% of the activity of the standard RPR-12535); however, dendrimer FIM-2126 displayed a substantially higher transfection efficiency as compared to the standard RPR-120535.

We acknowledge that all the products disclosed no significant toxicity at any charge ratio as compared to previously developed synthetic agents (similar levels as shown for FIM-2126). We conclude that the introduction of the degradable disulfides along polycationic peptide chains brings about a substantial reduction in toxicity at any charge ratio and for any of the disulfide containing polypeptides, however the complexes displayed moderate transfection probably due to early release of the DNA from the complex upon cell entry. On the other hand, the dendrimer product displayed substantial transfection efficiency. We have measured the size of some of the complexes used for transfection, and, interestingly, the complex dendrimer/DNA disclosed a size of about 20–40 nm, while the polycationic peptides displayed sizes of about 200–800 nm. We conclude that polycationic peptides should be long enough so that DNA can be released at a slower rate [37].

1.5 Towards Non-Electrostatic DNA Complexing Agents

Cationic lipids, cationic polymers, and cationic peptides (histone sequences) form compacted complexes with DNA. Along the different approaches, many advantages have been demonstrated for a variety of cationic entities, linkers, and lipids.

Nevertheless, the possibility exists to complex DNA through proton donor and proton acceptor non-electrostatic interactions with DNA-grooves. This type of non-electrostatic complexes has been developed especially for sequence-specific binding to DNA and not for gene delivery. Several prominent works were recently accomplished in this field [38–40].

The necessity of more efficient synthetic gene delivery methods prompted us to search for novel, less charged or non-cationic gene delivery systems. Such systems would thus not be based on electrostatic interactions with DNA. These non-electrostatic complexes can be advantageous for *in vitro* and *in vivo* applications, since unlike cationic lipid/DNA complexes, the novel molecules could not lead to a compacted state of DNA and therefore could potentially lead to different kinetics of DNA release from complexes. Several compounds are able to bind to double stranded DNA along the grooves by formation of hydrogen bonds. Groove binding typically exerts only subtle changes in DNA conformation, and DNA remains essentially in the native form. Therefore, groove-binding complexes will be essentially different from cationic lipid/DNA complexes, in which DNA is compacted within a multi-lamellar bilayer. We have set a prototype of new DNA vectors consisting of an amphiphile able to bind to DNA through hydrogen bond interactions based on Hoechst 33258 (**1**) (see Fig. 1.20), a well-known minor groove DNA binding agent [41]. Thus, alkyl derivatives of Hoechst 33258 were synthesized, and their complexes with DNA were characterized using physico-chemical methods and included comparative studies using known cationic lipid/DNA complexes for gene delivery. The synthesis of dodecyl (**2**) and octadecyl (**3**) carbamate derivatives of **1** are shown in Fig. 1.20.

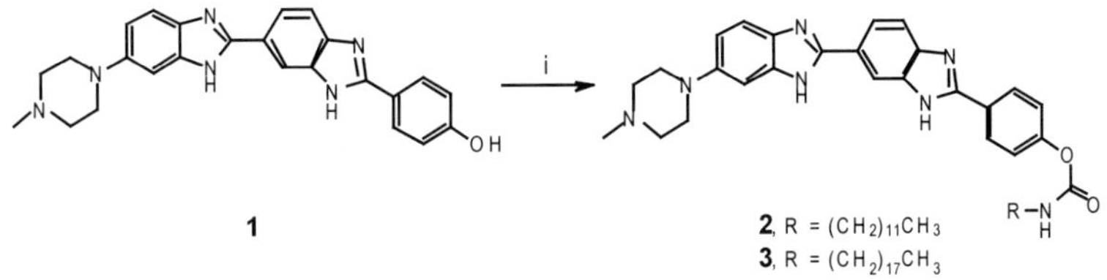

Figure 1.20 Non-electrostatic DNA binding lipids.

The physico-chemical characterization of complexes formed between plasmid DNA and products 2 and 3 demonstrates a different behavior of the new lipid/DNA complexes, as compared to previously described cationic lipids, that typically compact DNA and retard DNA in gel electrophoresis. The different nature of the presented complexes and especially their unperturbed electrophoretical mobility we have observed propose them as potential self-assembling systems for gene delivery [42].

1.6 Site-Specific Chemical Ligation of Targeting Peptides to Plasmid DNA

The SV40 large T antigen nuclear localization signal (NLS) efficiently transports proteins to the nucleus. We have developed a strategy for covalent site specific coupling of one or a controlled number of NLS peptides to plasmid DNA by triple helix formation. A chimera containing a psoralen-oligonucleotide-NLS peptide was synthesized. This conjugate was used to covalently associate one NLS peptide to plasmid DNA by triple helix formation and photoactivation. In similar studies, different probes were covalently linked to plasmid DNA without interfering with the structural and functional integrity of the plasmid [43, 44].

1.7 Concluding Remarks and Future Directions

The different non-viral self-assembling systems developed during more than 10 years research explore the nature of the cationic entity, the type of lipid, the geometry of the cationic entity, the biodegradability properties of linking functions, the controlled release of DNA from complexes and have introduced some elements for targeting DNA complexes. A great effort was invested to elucidate the physico-chemical properties of the different complexes. It seems most likely that extensive physico-chemical characterization of complexes formed between DNA and synthetic lipids, and their biological activity studies, becomes generally accepted. For the next few years, it is important to develop efficiently tissue-targeted self-assemblies taking advantage of the increasing number of tissue targeting molecules emerging, for example, from phage libraries [45, 46] and other targeting ligand technologies. Other exciting emerging nanotechnologies are composite nanoparticles able to pack genetic material in an internal cationic cavity of a synthetic polymeric particle, while the outer layer is composed of a cancer targeting polymer that can mediate the particle internalization and penetration to the nucleus and there releasing si-RNA [47]. Another very recent approach to be considered is the formation of nanoparticles formed exclusively from ds-DNA by a sonication fabrication process. These particles can be formulated with the classical non-viral gene delivery vectors or directly applied for gene delivery and expression [48]. In the meantime, non-viral self-assemblies with DNA remain a complementary approach to viral gene delivery rather than an independent competitive technology.

Acknowledgments

The research described here was supported by the Bioavenir program of the French Ministry of Industry, Rhone Poulenc-Rorer (Aventis), Centre National de la Recherche Scientifique (CNRS) and the Israel Science Foundation grant No. 274/02-4, and the Marcus Center for Medicinal Chemistry.

References

1. McPherson, J. D., Marra, M., Hillier, L., Waterston, R. H., Chinwalla, A., Wallis, J., *et al.* (2001). A physical map of the human genome. *Nature,* **409**, 934–941.
2. International Human Genome Sequencing Consortium (2001). Initial sequencing and analysis of the human genome. *Nature,* **409**, 860–921.
3. Boulikas, T. (1996). Gene therapy to human diseases: *ex vivo* and *in vivo* studies. *Int. J. Oncol.,* **9**, 1239–1251.
4. Smith, A. (1995). Viral vectors in gene therapy. *Nature,* **49**, 807–838.
5. French Anderson, W. (1998). Human gene therapy. *Nature,* **392**, 25–30.
6. Bischop, J. M. (1987). The molecular genetics of cancer. *Science,* **235**, 305–311.
7. Yang, Y., Nunes, F. A., Berencsi, K., Furth, E. E., Gonczol, E., and Wilson, J. M. (1994). Cellular immunity to viral antigens limits E1-deleted adenoviruses for gene therapy. *Proc. Natl. Acad. Sci. U.S.A.,* **91**, 4407–4411.
8. Fasbender, A., Lee, J. H., Walters, R. W., Moninger, T. O., Zabner, J., and Welsh, M. J. (1997). Complexes of adenovirus with polycationic polymers and cationic lipids increase the efficiency of gene transfer *in vitro* and *in vivo*. *J. Biol. Chem.,* **272**, 6479–6489.
9. Meunier-Durmort, C., Ferry, N., Hainque, B., Delattre, J., and Forest, C. (1996). Efficient transfer of regulated genes in adipocytes and hepatoma cells by the combination of liposomes and replication-deficient adenovirus. *Eur. J. Biochem.,* **237**, 660–667.
10. Byk, T., Haddada, H., Vainchenker, W., and Louache, F. (1998). Lipofectamine and related cationic lipids strongly improve adenoviral

infection efficiency of primitive human hematopoietic cells. *Hum. Gene Ther.*, **9**, 2493–2502

11. Schleef, R. R., Olman, M. A., Miles, L. A., and Chuang, J. L. (2001). Modulating the fibrinolytic system of peripheral blood mononuclear cells with adenovirus. *Hum. Gene Ther.*, **12**, 439–445.
12. Embinder, T., Sufaro, Y., Yusim, I., Byk, G., Passlick-Deetjen, P., Chaimovitz, C., *et al.* (2003). Correction of anemia in uremic mice by genetically modified peritoneal mesothelial cells. *Kidney Int.*, **63**, 2103–2112.
13. Themis, M., Forbes, S. J., Chan, L., Cooper, R. G., Etheridge, C. J., Miller, A. D., *et al.* (1998). Enhanced *in vitro* and *in vivo* gene delivery using cationic agent complexed retrovirus vectors. *Gene Ther.*, **5**, 1180–1186.
14. Sharma, S., Miyanohara, A., and Friedmann, T. (2000). Separable mechanisms of attachment and cell uptake during retrovirus infection. *J. Virol.*, **74**, 10790–10795.
15. Song, J. J., Kim, J., Lee, H., Kim, E., Kim, J., Park, Y. S., *et al.* (2000). Enhancement of gene transfer efficiency into human cancer cells by modification of retroviral vectors and addition of chemicals. *Oncol. Rep.*, **7**, 119–124.
16. Noguiez-Hellin, P., Robert-Le Meur, M., Salzmann, J. L., and Klatzmann, D. (1996). Plasmoviruses: nonviral/viral vectors for gene therapy. *Proc. Natl. Acad. Sci. U.S.A.*, **93**, 4175–4180.
17. Morozov, V. A., Noguiez-Hellin, P., Laune, S., Tamboise, E., Salzmann, J. L., and Klatzmann, D. (1997). Plasmovirus: replication cycle of a *novel nonviral/viral vector* for gene transfer. *Cancer Gene Ther.*, **4**, 286–293.
18. Fire, A., Xu, S., Montgomery, M., Kostas, S., Driver, S., and Mello, C. (1998). Potent and specific genetic interference by double-stranded RNA in Caenorhabditis elegans. *Nature*, **391**(6669), 806–811.
19. Koltover, I., Salditt, T., Radler, J. O., and Safinya, C. R. (1998). An inverted hexagonal phase of cationic liposome-DNA complexes related to DNA release and delivery. *Science*, **281**, 78–81.
20. Byk, G., Dubertret, C., Escriou, V., Frederic, M., Jaslin, G., Rangara, R., *et al.* (1998). Synthesis, activity, and structure–activity relationship studies of novel cationic lipids for DNA transfer. *J. Med. Chem.*, **41**, 224–235.
21. Byk, G., Soto, J., Mattler, C., Frederic, M., and Scherman, D. (1998). Novel Non-viral vectors for gene delivery: synthesis of a second generation library of mono-functionalized poly-(guanidinium)amines. *Biotech. Bioeng. (Comb. Chem.)*, **61**, 81–87.

22. Byk, G., Dubertret, C., Pitard, B., and Scherman, D. (1998). Transfecting compositions comprising DNA-binding, disulfide bond-containing compounds and their use in gene therapy. *Patent Application,* WO990128.

23. Byk, G., Scherman, D., Schwartz, B., and Dubertret, C. (1997). Lipopolyamines as transfection agents and pharmaceutical uses thereof. *Patent Application,* WO961774.

24. Byk, G., Frederic, M., and Scherman, D. (1997). One pot synthesis of unsymmetrically functionalized polyamines by a solid phase strategy starting from their symmetrical polyamine-counterparts. *Tetrahedron Lett.,* **38**, 3219–3222.

25. Byk, G., Dubertret, C., Schwartz, B., Frederic, M., Jaslin, G., Rangara, R., *et al.* (1997). Novel nonviral vectors for gene delivery: synthesis and applications. *Lett. Pep. Sci.,* **4**, 263–268.

26. Lee, E. R., Marshall, J., Siegel, C. S., Jiang, C., Yew, N. S., Nichols, M. R., *et al.* (1996). Detailed analysis of structures and formulations of cationic lipids for efficient gene transfer to the lung. *Hum. Gene Ther.,* **7**, 1701–1717.

27. Escriou, V., Ciolina, C., Lacroix, F., Byk, G., Scherman, D., and Wils, P. (1998). Cationic lipid-mediated gene transfer: effect of serum on cellular uptake and intracellular fate of lipopolyamine/DNA complexes. *Biochim. Biophys. Acta – Biomembranes,* **1368**, 276–288.

28. Pitard, B., Aguerre, O., Airiau, M., Lachages, A. M., Bouknikachvilli, T., Byk, G., *et al.* (1997). Virus-sized self assembling lamellar complexes between DNA and cationic micelles promote gene transfer. *Proc. Natl. Acad. Sci. U.S.A.,* **94**, 14412–14417.

29. Raspaud, E., Pitard, B., Durand, D., Aguerre-Chariol, O., Pelta, J., Byk, G., *et al.* (2001). Polymorphism of DNA/multi-cationic lipid complexes driven by temperature and salts. *J. Phys. Chem. B,* **105**(22), 5291–5297.

30. Carriere, M., Tranchant, I., Niore, P. A., Byk, G., Mignet, N., Escriou, V., *et al.* (2002). Optimization of cationic lipid mediated gene transfer: structure-function, physico-chemical, and cellular studies. *J. Liposome Res.,* **12**(1 & 2), 95–106.

31. Leclercq, F., Cohen-Ohana, M., Mignet, N., Sbarbati, A., Herscovici. J., Scherman, D., *et al.* (2003). Design, synthesis and evaluation of gadolinium cationic lipids as tools for biodistribution studies of gene delivery complexes. *Bioconjugate Chem.,* **14**, 112–119

32. Schwartz, B., Ivanov, M-A., Pitard, B., Escriou, V., Rangara, R., Byk, G., *et al.* (1999). Synthetic DNA-compacting peptides derived from human

sequence enhance cationic lipid-mediated gene transfer *in vitro* and *in vivo*. *Gene Ther.*, **6**(2), 282–292.

33. Frederic, M., Scherman, D., and Byk, G. (2000). Introduction of cyclic guanidines into cationic lipids for non-viral gene delivery. *Tetrahedron Lett.*, **41**(5), 675–679.

34. Byk, G., Wetzer, B., Frederic, M., Dubertret, C., Pitard, B., Jaslin, G., *et al.* (2000). Reduction-sensitive lipopolyamines as a novel nonviral gene delivery system for modulated release of DNA with improved transgene expression. *J. Med. Chem.*, **43**, 4377–4387.

35. Wetzer, B., Byk, G., Frederic, M., Airiau, M., Blanche, F., Pitard, B., *et al.* (2001). Reducible cationic lipids for gene transfer. *Biochem. J.*, **356**(3), 747–756.

36. Mignet, N., Byk, G., Wetzer, B., and Scherman, D. (2003). DNA complexes with reducible cationic lipid for gene transfer. *Methods Enzymol.*, **373**, 357–369.

37. Mirkin, F., Eini, H., Douvdevani, A., and Byk G. (2009). New degradable cationic peptides for modulated gene delivery. *Adv. Exp. Med. Biol.*, **611**, 245–246.

38. Dervan, P. B.(1997). Gene-specific transcription inhibition *in vivo* by designed ligands. *FASEB J.*, **11**, 2546.

39. Kielkopf, C. L., White, S., Szewczyk, J. W., Turner, J. M., Baird, E. E., Dervan, P. B., *et al.* (1998). A structural basis for recognition of A center dot T and T center dot A base pairs in the minor groove of B-DNA. *Science*, **282**, 111–115.

40. Minehan, T. G., Gottwald, K., and Dervan, P. B. (2000). Molecular recognition of DNA by Hoechst benzimidazoles: exploring beyond the pyrrole-imidazole-hydroxypyrrole polyamide-pairing code. *Helv. Chim. Acta*, **83**, 2197–2213.

41. Soto, J., Bessodes, M., Pitard, B., Mailhe, P., Scherman, D., and Byk, G. (2000). Non-electrostatic complexes with DNA: towards novel synthetic gene delivery systems. *Bioorg. Med. Chem. Lett.* **10**, 911–914.

42. Scherman, D., Bessodes, M., Pitard, B., Soto, J., and Byk, G. (2001). Oligobenzimidazole derivatives and their use as DNA transfection agents. *Patent Application,* WO 0132630.

43. Ciolina, C., Byk, G., Blanche, F., Thuillier, V., Scherman, D., and Wils, P. (1999). Coupling of nuclear localization signals to plasmid DNA and specific interaction of the conjugates with importin. *Bioconjugate Chem.*, **10**, 49–55.

44. Neves, C., Byk, G., Scherman, D., and Wils, P. (1999). Coupling of a targeting peptide to plasmid DNA by covalent triple helix formation. *FEBS Lett.*, **453**, 41–45.

45. Ruoslahti, E. (1996). RGD and other recongnition sequences for integrins. *Annual Rev. Cell Develop. Biol.*, **12**, 697–715.

46. Pasqualini, R., Koivunen, E., and Ruoslahti, E. (1996). Alpha integrins as receptors for tumor targeting by circulating ligands. *Nature Biotech.*, **15**, 542–515.

47. Blackburn, W. H., Dickerson, E. B., Smith, M. H., McDonald, J. F., and Lyon, L. A. (2009). Peptide-functionalized nanogels for targeted siRNA delivery. *Bioconjugate Chem.*, **20**, 960–968.

48. Shimanovich, U., Eliaz, D., Adva Aizer, A., Irena Vayman, I., Shulamit Michaeli, S., Shav-Tal, Y., *et al.* (2011). Sonochemical synthesis of DNA nanospheres. *ChemBioChem*, **12**, 1678–1681.

Chapter 2

Cationic Polymers for the Delivery of Therapeutic Nucleotides

Wahid Khan, Saravanan Muthupandian, and Abraham J. Domb*

School of Pharmacy, Faculty of Medicine, The Hebrew University of Jerusalem, Jerusalem 91120, Israel

*avid@ekmd.huji.ac.il

Gene therapy is a promising approach for treatment or prevention of wide range of diseases associated with defective gene expression. It involves the insertion of a therapeutic gene into cells, followed by expression and production of the required proteins. Two main delivery systems, including viral or non-viral gene carriers, are currently deployed for gene therapy. Although viruses are currently the most commonly researched vector, however, the need to resolve drawbacks related to viral vectors (e.g., high risk of mutagenicity, immunogenicity, low production yield, limited gene size) led to the development of non-viral vectors.

Several non-viral modalities are reported to transfer foreign genetic material into cells; cationic polymers constitute one of the most promising approach. For this reason, various cationic polymers with diminished cytotoxicity and enhanced efficacy are rapidly emerging as systems of choice. This chapter provides an overview and recent developments of cationic polymers employed for *in vitro*

Nanotechnology for the Delivery of Therapeutic Nucleic Acids
Edited by Dan Peer

ISBN 978-981-4411-04-2 (Hardcover), 978-981-4411-05-9 (eBook)
www.panstanford.com

and *in vivo* delivery of therapeutically important nucleotides, e.g., DNA and siRNA. The mechanism and recent progress in cationic polymer based gene delivery are reviewed in detail.

2.1 Introduction

The basic concept of gene therapy is that disease can be treated by transfer of genetic material into specific cells of a patient to supplement defective genes responsible for disease development. Stribley *et al.* demonstrated two strategies involved for the application of gene delivery: (i) Corrective therapy — correction of genetic defects in target cells — is exploited for the treatment of diseases with single gene disorders like severe combined immunodeficiency syndromes, cystic fibrosis, hemophilia, sickle cell anemia, β-thalassemia, muscular dystrophy and lignant tumors, including ovarian carcinoma, and (ii) cytotoxic gene therapy — destruction of target cells using a cytotoxic pathway; it is used for the treatment of malignant tumors, including ovarian, breast, and endometrial carcinomas.[1]

The fundamental idea is to deliver the gene to cells or tissues. It may be activation, silencing, introducing or gene knock out and knock down both *in vitro* and *in vivo* toward cells, and the possibilities for interventions.[2] Successful gene therapies depend on the efficient delivery of genetic materials into the cell nucleus and its effective expression within these cells. DNA can be delivered into the cell nucleus either using physical means or by specific carriers that carry the genes into the cells. Number of techniques has been developed for DNA delivery, including direct introduction of transgene using cell electroporation, microinjection of DNA, and incorporation of the gene by vectors.[3]

Vectors for delivering genes can be divided into two main groups: (a) Viral carriers, where the DNA to be delivered is inserted into a virus, and (b) cationic molecular carriers, which forms electrostatic interactions with DNA for delivering gene to cells and includes polymers and lipids.[4] Viral vectors, including retroviruses, adenoviruses, and adeno-associated viruses, are effectively used for introducing genetic material into host cells,[5] but immunogenicity, inflammatory effects, and safety concerns with use of these viruses restrict their usefulness.

Although viruses are currently the most commonly researched vector, because of continuing safety concerns research has broadened for developing non-viral alternatives.[6] Non-viral vectors have advantages that they do not integrate into chromosome, can introduce DNA into non-dividing cells, have low immunogenicity, have ability to deliver large genes, have no infective risk, and are significantly less expensive, easy to handle, and, most importantly, have possibility for large-scale production at reasonable cost.[7] Non-viral systems offer several advantages, but they are also associated with few major disadvantages, including toxicity and low transfection efficiency, since the non-viral delivery vectors have to overcome intracellular barriers, such as endosomes and nuclear membranes.[8] Thus, the most challenging task is to design non-viral gene delivery vectors with low cytotoxicity and high transfection efficiency. Non-viral molecular carriers include polycation (mainly ion-based/cationic polymers and lipids delivery system) to assist the passage of DNA to the cell's nucleus.[6]

In the late 1950s, the earliest cationic polymer vector for gene delivery was originated to enhance nucleic acid to enter the cells. Thereafter, in 1965, dextran was used in gene delivery as the simple, effective, and widely used polymer.[9] Cationic polymers were intensively studied as non-viral gene vectors since the first successful clinical gene therapy of adenosine deaminase deficiency in 1989.[10] Cationic vectors with diminished cytotoxicity and enhanced efficacy are rapidly emerging as systems of choice. Presently, polycations used as gene delivery systems are mostly polyamines that contain primary, secondary, or tertiary amines. Major cationic polymers that are commonly used in gene delivery include poly(L-lysine) (PLL), polyethyleneimine (PEI), dendrimers, cationic polysaccharides, etc.[11–13]

2.2 Cationic Polymer Targeted Delivery of Nucleotides

Delivery of therapeutic nucleotides to the target is possible through various methods and mechanisms. Cationic polymers are leading class of non-viral gene delivery systems due to their versatility in MW, polymer type, molecular architecture, and ability to introduce target-specific moieties. The degradation of these polymers releases plasmid DNA (pDNA) into the cytosol, and these polymers have been found to be better candidates for gene transfers rather than lipid-based gene carriers.[14]

The significant advantage of cationic polymeric delivery systems is that they can easily complexed with the anionic DNA molecules. The mechanism of polymer-DNA-complex (polyplexe) formation includes a general phenomenon where electrostatic interaction of cationic polymers with anionic DNA gives rise to polyplexes (Fig. 2.1). These cationic polyplexes then interact with the negatively charged cell surface resulting in improved DNA uptake in cell via endocytosis.[9] After cellular entry, these polyplexes undergo dissociation via endosomal escape to release nucleic acid into the nucleus for gene expression.[15] Additionally, cationic polymers have ability to condense large DNA to smaller structures;[16] these polyplexes have tightly packed DNA, thus the entrapped DNA is protected from DNase.[17] One of the frequently used methods for preparing polyplexes includes direct mixing method, in which DNA and aqueous suspensions of cationic polymers are mixed together, which gives condensed DNA enclosed by a cationic bilayer with a required particle diameter.[18]

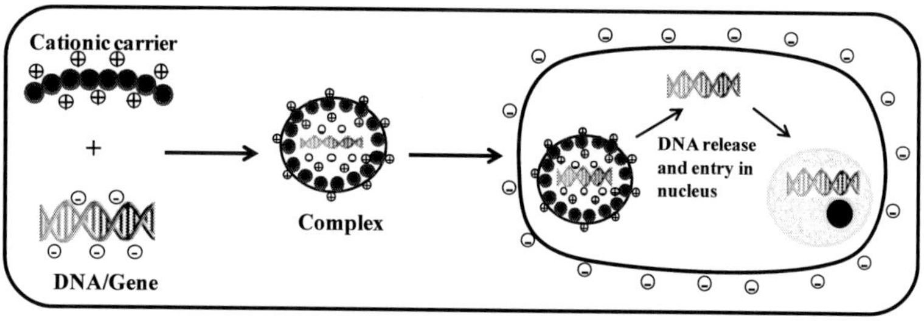

Figure 2.1 Mechanism for targeted delivery of nucleotides using cationic polymer carrier.

Chitosan, PEI, PLL are some of the cationic polymers that can also be formulated to nanoparticulate delivery systems, and these nanoparticles help efficiently to deliver DNA with or without any therapeutically active agents to cells. These polymeric nanoparticles are stable while delivering DNA to the target, but when they are taken up by the target cells, they destabilize themselves quickly to liberate the DNA. For most successful intracellular drug delivery, polyplexes/nanoparticles prepared must have a diameter of <100 nm.[19]

2.3 Major Cationic Polymers Used for Delivery of Nucleotides

Gene therapy holds great promise for treating a variety of human diseases and conditions. The field of gene therapy has advanced rapidly in the last two to three decades. However, a major limiting factor remains the lack of a suitable gene delivery vector. The passage of the transgene through the cell to the nucleus is hampered by many obstacles. The ideal vector must be stable in the systemic circulation, escape the reticuloendothelial system, be able to extravasate tissues, enter the target cell, escape lysosomal degradation, and transport DNA to the nucleus to be transcribed. With increasing understanding of the physicochemical properties essential to overcome various barriers, it is possible to apply rational design to the cationic carriers.[20] Thus, various approaches to maximize gene expression are presently under investigation,[6] and a number of cationic polymers, block co-polymers, dendrimers, etc., have been rationally designed to optimize gene delivery.

2.3.1 Polyethylenimine

Polyethylenimine (PEI) is a branched, linear or dendrimer polyamine polymer (Fig. 2.2), principally applied in gene transfection since 1995. It is simple and easy to prepare and to modify and relatively safe compared to viral vectors.[21] It is composed of 25% primary, 50% secondary, and 25% tertiary amines with high buffering capacity. It exhibits proficient endosomal release and contains secondary and tertiary amines exhibiting pKa values between physiological and lysosomal pH. PEI has an advantage over other polycations that it exhibits strong DNA compaction capacity with an intrinsic endosomolytic activity. It has been termed the "proton sponge" polymer, and this hypothesis is used to explain endosomal disruption by cationic polymers with ionizable amine groups. The "proton sponge" nature of PEI leads to buffering inside endosomes.[22]

PEI based non-viral vectors have been delivered both locally or systemically to target gene to tumor, lung, liver, etc. Although this requires strategies to efficiently shield transfection polyplexes against non-specific interaction with blood components, extracellular matrix and untargeted cells and the attachment of targeting moieties allow for the directed gene delivery to the desired cell or tissue.[23] Owing to high transfection efficiency, branched PEI with a molecular weight (MW) of 25 kDa is considered to be the "gold standard" for polymeric non-viral vectors.[24] High transfection efficiency of PEI, along with its cytotoxicity, strongly depends on its MW.

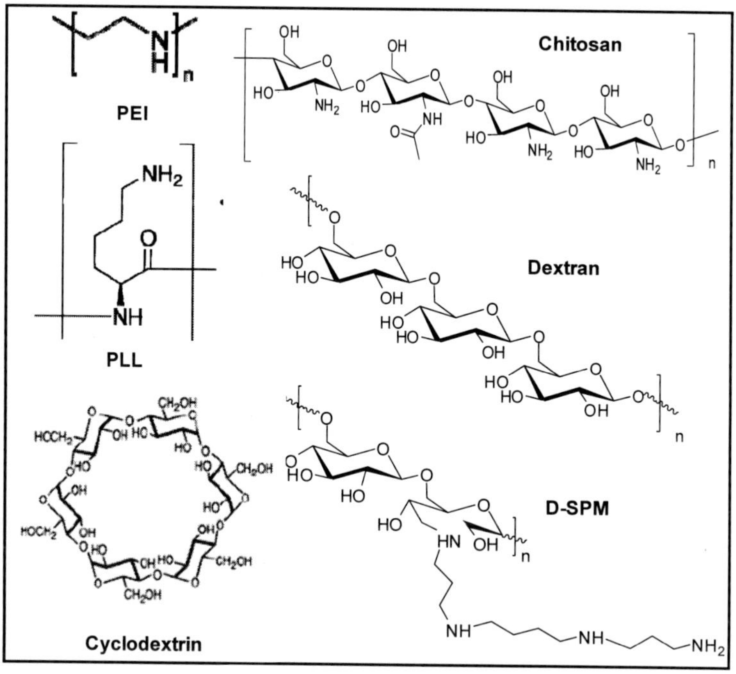

Figure 2.2 Chemical structure of cationic polymers for the delivery of nucleotides.

To enhance its gene delivery efficiency and minimize cytotoxicity, small cross-linked PEI with biodegradable linkages were synthesized and evaluated for their transfection efficiencies *in vitro*.[25] Chen *et al.* reported a simple, reproducible process to condense pDNA with PEI. PEI-DNA complexes with 30–60 nm diameter and stable for more than one month at 4°C with respect to size and transfection activity were prepared. Intraperitoneal administration of these PEI-condensed DNA resulted in significant gene expression in a human ovarian cancer peritoneal xenograft model.[26] Cherng *et al.* reported a simple three blending methods to combine biodegradable polyurethane and PEI with DNA by unique mixing sequences. The complex prepared showed better transfection efficiency than PEI alone and also coupled with better biocompatibility and no cytotoxicity.[27]

Linear PEI (LPEI) with a MW of 22 kDa has superior ability to induce gene transfer compared to its branched form. However, the transfection efficiency of the polymer cannot be enhanced beyond a certain limit due to cytotoxicity. Breunig *et al.* explored LPEI with MW ranging from 1.0 to 9.5 kDa to overcome this limitation. Results suggested that LPEI with low MW are promising candidates for non-viral gene delivery. They are more efficient and substantially less toxic than their higher MW counterparts.[28] Liu *et al.* reported low MW PEI (LMW-PEI) linked to an expressing plasmid containing a green fluorescence protein reporter gene for effective gene transfer in CM7221 cell line. Relationship among the MW, structure of PEI, their transfection activity, and cytotoxicity was studied. Results showed that LMW-PEI/DNA complexes led to high levels (about 55%) of expression in the CM7221 cell line. However, with the increasing of PEI MW, the transfection activity of PEI was decreasing and simultaneously there was an increasing cytotoxicity with these larger PEI molecules.[29]

Li *et al.* studied the influences of plasmids factor, albumin, serum, cell density, operation methods, and PEI/DNA preserve factors on transfection efficiency. Linear plasmids lowered transfection efficiency. Serum and albumin in the culture medium decreased the transfection efficiency of PEI/DNA. Freezing of PEI/DNA complexes significantly decreased transfection efficiency. Thus, with PEI, optimal and reduplicate transfection results could be obtained by controlling related factors.[30] Goyal *et al.* reported LPEI (25 kDa) nanoparticles (LPN) series cross-linked with 1,4-butanediol diglycidyl ether. Size, surface charge, morphology, pDNA protection/release, cytotoxicity, and transfection efficiency of LPN was studied. These LPN showed increased buffering capacity with increasing cross-linking and also exhibited excellent transfection efficiency in comparison with LPEI and the commercial transfection agents. These particles were relatively non-toxic *in vitro* (in cell lines) and *in vivo* (in Drosophila). *In vivo* gene expression studies in Balb/c mice showed maximum expression of the reporter gene in the spleen.[31]

Synthesized PEI derivatives are more advantageous over conventional PEI for gene delivery. Forrest *et al.* reported ester cross-linked PEI derivatives prepared by reacting PEI (800 Da) with small diacrylates mediates gene expression more efficiently than PEI

(25 kDa). In addition, these PEI derivatives did not have cell toxicity, and the ester bonds were susceptible to hydrolysis.[32] Similarly, to enhance the gene delivery efficiency and decrease cytotoxicity of polyplexes, copolymers consisting of branched PEI (25 kDa) grafted with Pluronic (F127, F68, P105) were successfully synthesized by Liang *et al.* The copolymers were tested for cytotoxicity, DNA condensation, and complexation properties. Their polyplexes with pDNA were characterized in terms of DNA size, surface charge, and transfection efficiency. The complexes were of sizes below 300 nm, which implicated their potential for intracellular delivery. The complex exhibited better condensation and complexation properties than PEI 25 kDa, and cytotoxicity of PEI was strongly reduced after copolymerization. These results demonstrate that polyplexes prepared provided promising properties as stable, high transfection efficiency vectors.[33]

A series of PEI and γ-polyglutamic acid (PGA) nanocomposites (PPGA) were prepared and evaluated by Tripathi *et al.* On complexion with pDNA, the positively charged PPGA/DNA nanocomposites resulted in a higher level of *in vitro* reporter gene transfection as compared to native PEI and selected commercial reagents. These nanocomposites also delivered siRNA efficiently into mammalian cells. These results demonstrated the potential of the reported nanocomposites for *in vivo* gene delivery.[34] Similarly, low MW PEI was grafted onto the block copolymer of PLL and PEG, yielding a ternary copolymer PEG-b-PLL-g-PEI for gene delivery. Results indicate that the copolymers to be a promising vector combining low cytotoxicity, biodegradability, and high gene transfection efficiency.[35]

Branched PEI is capable of forming self-assembly complexes with DNA to become a highly efficient agent for use in gene delivery. Conjugation through the primary amines of PEI is an approach to further enable the targeted delivery or to improve the stability of the DNA-polymer complexes. Tseng *et al.* investigated the effects of conjugation, folate, and the dextrans of MW 1500 and 10000 to prepare three different types of PEI conjugates: dextran-PEI, folate-PEI, and folate-dextran-PEI to form complexes with DNA. These conjugates were able to deliver an estimated amount plasmid into the cells and found to cause less cytotoxicity than the unmodified PEI.[36] Similarly, with the aim to improve the specificity and to reduce the cytotoxicity of PEI, Lee *et al.* synthesized the conjugates of the branched PEI (25 kDa) with transferrin (TF), a cell-binding ligand. The transferrin-PEI (TP) conjugates were synthesized, and transfection efficiency of TP conjugates was evaluated. From the results, TP conjugates showed lower cytotoxicity compared with the PEI system.[37] Further, to improve specificity and reduce its cytotoxicity, PEI was modified by PEG to reduce the protein interaction and cell cytotoxicity of delivery vectors. Thus, PEG-TF-PEI conjugate as an efficient and safe carrier of pDNA was synthesized. Nanocomplexes of conjugates with pDNA were characterized by measuring the particle size and the surface charge. It was found that transfection efficiency of nanocomplexes in Jurkat cells was improved and cytotoxicity was decreased compared with PEI complex.[38]

Recently, Fortune *et al.* explored the effect of increased hydrophobicity of the PEI on the efficacy and specificity of gene delivery. N-alkylated LPEI with varying alkyl chain lengths and extent of substitution was synthesized and characterized, and *in vivo* transfection efficiency, specificity, and biodistribution were investigated. N-propyl-PEI was found best in the liver and heart, while N-octyl-PEI improved expression in the kidneys over the parent PEI polymer. These enhancements in gene expression occur without changing the plasmid biodistribution. Thus, alkylation does not alter the cellular uptake but enhances transfection subsequent to cellular uptake.[39] Further, gold nanoparticles conjugated with low MW PEI (800 Da) were synthesized (GNP-PEI) and characterized as gene transfection vectors by Hu *et al.* These complexes showed high efficiency for gene delivery in monkey kidney cells *in vitro*. GNP-PEI exhibited transfection efficiency exceeding 60 fold that of PEI (25 kDa) in 10% serum medium. All GNP-PEI exhibited mild cytotoxicity in comparison with that of PEI 25 kDa.[40] Similarly, Sharma *et al.* examined the gene transfer efficiency and toxicity of 2 kDa PEI conjugated to gold nanoparticles in the human cornea *in vitro* and rabbit cornea *in vivo*. The results suggest that PEI-GNP are safe for the cornea and can potentially be useful for corneal gene therapy.[41]

2.3.2 Poly(L-lysine)

PLL is one of the first polymers used for DNA condensation. PLL has a sufficient number of primary amines with positive charges to interact with the negatively charged phosphate groups of DNA. The complex formation of PLL and DNA has lot of constraints in clinical applications due to relatively low competency. Low cytotoxicity and high gene transfection efficiency are critical issues in designing non-viral gene delivery vectors. Therefore, PLL are modified primarily at the reactive primary amino groups of the lysine residues as side chains with various functional hydrophobic groups such as thiol and histidyl moieties.[42] To improve aqueous solubility of PLL/DNA complexes, reductive amination of amino groups of PLL and the reductive ends of hydrophilic dextran were reported.[43, 44] Accordingly, PLL was glycosylated using methylglycolate to decrease cytotoxicity.[45] Choi *et al.* reported covalent linking PLL to PEG to decrease both cytotoxicity and aggregation of PLL.[46]

To produce reducible, effective biocompatible, and non-toxic gene delivery system, Nounou *et al.* designed and synthesized novel reducible linear L-lysine modified copolymers (LLC). The transfection efficiencies of the polyplexes showed that LLC polyplexes produced five times higher transfection efficiencies in HDF cells, three times higher transfection efficiencies in MCF-7 cells, and four times higher transfection efficiencies in MA cells as compared to the optimal PLL control. The LLC/pDNA polyplexes showed significantly lower cytotoxicity in HDF, MCF-7, and MA cells. These results suggest that these LLC are efficient, reducible, and biocompatible polymers for gene delivery.[47]

Fu *et al.* synthesized novel biodegradable poly(lactic acid)-poly(ethylene glycol)-poly(L-lysine) (PLA-PEG-PLL) copolymer, and explored its applicability and feasibility as a non-viral vector for gene transport. The physicochemical properties (morphology, particle size, and surface charge) and the biological properties (protection from nuclease degradation, plasma stability, *in vitro* cytotoxicity, and *in vitro* transfection ability) of the gene-loaded PLA-PEG-PLL nanoparticles were evaluated and reported as very attractive candidate as a non-viral vector for gene delivery *in vivo*.[48]

Bone marrow stromal cells (BMSC) represent an important cell phenotype for pursuit of successful gene therapy. Farrell *et al.* compared PLL and branched PEI as gene carriers for modification of BMSC. Although both PLL and PEI were readily internalized with BMSC, but the relative uptake of PEI was significantly higher as compared to the PLL, but both PLL and PEI readily displayed cellular uptake.[49] Yamagata *et al.* investigated the relationships between structures of PLL molecules with respect to their gene delivery activities and compared dendritic PLL (KG6) with linear PLL. KG6 formed a neutral DNA complex, and its DNA compaction level was weaker than that of PLL. The amount of DNA binding and uptake into cells mediated by PLL was fourfold higher than that with KG6. However, KG6-mediated gene expression was 100 fold higher than by PLL. Amount of induced cytokine production after intravenous injection of DNA complexes with KG6 and PLL was low and was found similar to when DNA was injected alone.[50]

Zhang *et al.* designed replica particles based on PLL polymers cross-linked via a homobifunctional linker to support co-adsorption of pDNA and a peptide hormone for concurrent transfection and induction of a cellular function. PLL replica particles were prepared by infiltrating polymer into mesoporous silica particles, cross-linking the adsorbed chains by using a homobifunctional cross-linker and finally removing the template particles. Melanin stimulation and gene expression were simultaneously evaluated and PLL replica particles were found as a promising vector for gene therapy and hormone stimulation.[51] Similarly, a terplex gene delivery system consisting of pDNA, low density lipoprotein (LDL), and hydropholized PLL was designed and characterized by Kim. pDNA when formulated with hydropholized PLL and LDL forms a stable and hydrophobicity/charge-balanced terplex system for efficient cellular uptake. Terplex DNA complex showed significantly longer retention in the vascular space than naked DNA. This system was used in the augmentation of myocardial transfection at an infraction site with VEGF gene.[52]

2.3.3 Cationic Polysaccharides

Cationic polysaccharides are attractive candidates for gene delivery. They are non-toxic, biodegradable, and biocompatible materials. The use of biodegradable polysaccharide carriers is especially suitable for transfection and biological applications because they are water soluble, can be readily transported to cells *in vivo*, and thus act as an effective vehicles for transporting agents complexed with them.[53] Major cationic polysaccharides used in therapeutic purposes are natural (chitosan and cyclodextrin) or semisynthetic in origin. Semisynthetic cationic polysaccharides have been synthesized by conjugation of various oligoamines to oxidized polysaccharides by reductive amination and were tested for antiprion activity.

Polycations of dextran, pullulan, and arabinogalactan grafted with oligoamines of 2 to 4 amino groups were also investigated and were found effective in gene delivery.[54]

2.3.3.1 Chitosan

Chitosan is a cationic linear polymer obtained from chitin comprising copolymers of β(1-4)-glucosamine and randomly located N-acetyl-D-glucosamine. It has been proved to be biodegradable, bio-compatible, non-antigenic, non-toxic, and biofunctional polymer.[55] It can be fully or partially deacetylated form of chitin; the degree of deacetylation of typical commercial chitosan is usually between 70% and 95%, and the MW ranges between 10 and 1000 kDa.

Chitosan is a non-toxic polymer, degraded by lysozymes into common amino sugars. Being cationic and biodegradable in nature, it has been widely used as gene delivery vector.[56] Due to the high positive charge, it can easily form polyelectrolyte complexes with negatively charged nucleotides by electrostatic interactions.[57] NOVAFECT chitosans are ultrapure chitosan oligomers that were marketed as carriers for non-viral gene therapy. Klausner *et al.* have reported chitosan-DNA nanoparticles based on NOVAFECT.[58]

Chitosan both in native oligomer form as well as copolymer is used in gene delivery. *In vitro* transfection studies demonstrated the ability of oligomeric chitosan-DNA nanoparticles to effectively transfect COS-7 cells. Oligomeric chitosan-DNA nanoparticles were evaluated for the treatment of ocular diseases of the cornea and for various other applications. Looking to various roles of chitosan to play in siRNA delivery, a novel chitosan-based siRNA nanoparticle delivery system has been proposed.[59] To improve the transfection efficiency of chitosan, PEI (MW 1.8 kDa) was linked to chitosan to form a complex by Gao *et al.*[60] However, Weecharangsan *et al.* evaluated chitosan lactate (CL) as a DNA complexing agent, for its efficiency in transfecting green monkey fibroblasts, and effect on cell viability was compared with PEI. Results suggested advantage of CL for *in vitro* gene transfection, with the ease of preparation of polymer/DNA complexes and low cytotoxicity.[61]

Wang *et al.* prepared a series of chitosan-graft-PEI (chitosan-g-PEI) copolymers as gene carriers with high transfection efficiency and low cytotoxicity. Chitosan-g-PEI interacts efficiently with pDNA, and optimal transfection efficiency was observed with significantly lower toxicity compared with PEI (25 kDa). Toxicity increased with increasing MW of the PEI segment in chitosan-g-PEI.[62] Yu *et al.* prepared copolymers poly(L-lysine)-graft-chitosan (PLL-g-Chi). The gel retardation electrophoresis showed that the PLL-g-Chi copolymers possessed better pDNA-binding ability than chitosan. The gene transfection effect in HEK 293T cells of the copolymers was better than that of chitosan and was dependent on the PLL grafting ratio. The PLL-g-Chi copolymers could be used as an effective gene delivery vectors.[63]

2.3.3.2 Cyclodextrins

Cyclodextrins (CD) is non-toxic and non-immunogenic and is produced from starch by means of enzymatic conversion. It is cyclic oligomer of glucose units having a unique amphiphilic structure that consists of a hydrophobic cavity and hydrophilic exterior and thus building CD as water-soluble. CD contains a number of glucose monomers ranging from six to eight units in a ring, creating a cone shape and thus (a) α-cyclodextrin: six-membered sugar ring molecule, (b) β-cyclodextrin: seven sugar ring molecule, and (c) γ-cyclodextrin: eight sugar ring molecule.

CD forms inclusion complexes with small hydrophobic molecules and has been incorporated into a gene delivery polymer to provide a facile and versatile attachment site for targeting ligand. Forrest *et al.* derivatized PEI with β-CD (CD-PEI) on approximately 10% of the polymer's amines. Human insulin was derivatized with a hydrophobic palmitate group (pal-HI), which could anchor the protein to CD-PEI/DNA polyplexes. CD-PEI was non-toxic to HEK293 cells and mediated fourfold higher gene expression than unmodified PEI. Addition of the pal-HI to CD-PEI enhanced gene expression. CD-binding moieties are attached with relative ease to various types of ligands; thus CD-PEI may be a useful material for testing novel cell-specific targeting compounds.[64]

The influence of CD in forming grafting complexes showed that the more the grafting percentage of the CD, the less cytotoxic the polymer will be. Transfection efficiency also increased if the grafting percentage was increased.[65] Pun *et al.* reported linear and branched PEI (LPEI and bPEI), grafted with β-CD, and investigated them as *in vitro* and *in vivo* gene delivery agents. The *in vitro* toxicity and transfection efficiency are sensitive to the level of CD grafting. The CD containing polycations are combined with adamantane-

PEG (AD-PEG) conjugates to form particles, and gene expression observed was equal or greater than LPEI. Tail vein injections into mice of pDNA formulated with CD-LPEI and AD-PEG do not reveal observable toxicities, and both nucleic acid accumulation and expression are observed in liver.[66]

As non-viral gene delivery vectors, Yang *et al.* synthesized series of novel cationic star polymers by conjugating multiple oligoethylenimine arms onto α-CD core. Prepared polymers are found to be less toxic, biocompatible and do not elicit immune reactions.[67] Jiang *et al.* studied target ligand folic acid and cell-penetrating peptide octaarginine coupled with the gene vectors composed of β-CD and low MW PEI (600 Da) for highly efficient gene delivery to tumor cells. Study provides a promising means to produce gene nanovectors for *in vivo* applications.[68]

2.3.3.3 Dextran, dextran-spermine

Dextran is a biodegradable and biocompatible polymer. Its straight chain consists of α-1,6 glycosidic linkages with few α-1,3 glycosidic linkages as the branch linkage. Dextran is readily soluble in water and electrolyte solutions. It binds to erythrocytes and platelets, increasing their electronegativity and thus reducing erythrocyte aggregation. Dextran is biocompatible and used as plasma expanders and for various other uses.

Cationic dextrans were prepared with reagents carrying positive charge; conjugation through primary amines is one of the most commonly used methods. Cationization of dextran increases its gene delivery efficiency similar to that of PEI.[69, 70] Tseng *et al.* reported various degrees of grafting of dextran onto PEI in an effort to investigate how the conjugation affected the mechanisms of transgene expression. Grafting of dextran onto PEI was found to significantly diminish the cytotoxicity, buffering capacity, cellular entry, and the integrity of the DNA-polymer complexes. Results demonstrated that the conjugation of dextran onto the primary amines of PEI inhibited the entry of plasmid across the cell membrane.[71]

Cationic polysaccharides are also synthesized by conjugation of various oligoamines to oxidized polysaccharides via reductive amination.[72] More than 300 different polycations were prepared starting from various polysaccharides of different MW and oligoamines having two to four amine groups. Although most of these conjugates formed stable complexes with DNA, only the dextran-spermine (D-SPM) polycations of defined MW are found to be active in transfecting a wide range of cell lines *in vitro* attributed to spermine residues, which play a crucial role in cell transfection. The complexes formed between D-SPM and pDNA exhibit a hydrophilic surface due to the saccharides building units of the polymer backbone. Such hydrophilic nature limits the cell uptake of the complexes. Azzam *et al.* reported higher spermine ratios of grafted oligoamines resulted in high gene expression, whereas low to negligible expressions were obtained with lower spermine contents.[53]

Further, hydrophobization of D-SPM was achieved by treating the polymer with N-hydroxysuccinimide derivatives of cholesterol and fatty acids. The oleate-modified D-SPM remarkably enhanced the gene expression in contrast to unmodified D-SPM, which resulted with a drastic decrease in the transfection yields. Modified D-SPM derivatives of other fatty acids and cholesterol showed improved transfection yields in comparison to unmodified D-SPM. This improvement in cell transfection was attributed to oleate residues that probably play a role in increasing stability and uptake of polycation-DNA complexes. Similarly, Hosseinkhani *et al.* reported modification of the D-SPM polycation with PEG resulted in high transfection yield in serum-rich medium. This offers a new biodegradable polycation based on natural components, which is capable of transfecting cells and tissues *in vitro* and *in vivo*.[73]

2.3.4 Dendrimers

Dendrimers are spherical, highly branched, three-dimensional polymers having ability to interact with various forms of nucleic acids such as pDNA, antisense oligonucleotides, and RNA. The interaction between the dendrimer and the nucleic acids is purely electrostatic where the cationic dendrimer condenses the anionic nucleic acids. The nature of the dendrimer nucleic acid complex depends on various factors like stoichiometry, concentration of dendrimer-amines and nucleic acid-phosphates, as well as bulk solvent properties like pH, salt concentration, etc. The net positive charge of the dendrimer nucleic acid complex determines the transfection efficiency. Dendrimer helps in DNA condensation and endosome release due to the presence of cationic amino acid residues in the polymeric structure. Complexes formed protect the nucleic acid from degradation.[3, 74]

Dendrimers are prepared either by divergent (starting from a central core molecule) or convergent (starting with what will become the periphery of the molecule, building inwards) strategies. The degree of branching is expressed in the generation of the dendrimer. PAMAM Starburst dendrimers of generation G3 to G10 were found to form stable complexes with DNA, but their ability to transfect different cell lines varied considerably. Higher-generation dendrimers (G5 to G10) were found to be of superior efficiency.[74]

The structure of dendrimer shows high densities of amines in the periphery of the molecule. These outer amines enable efficient condensation of nucleic acids, leaving the inner amine functions available for a proton sponge effect during endolysosomal acidification, thus enabling more efficient endosomal escape. An increased flexibility of the polymer with a better ability to complex DNA might play a crucial role.[75, 76] Dendrimer delivery systems are useful and less toxic because they can be modified *in vivo* to have defined molecular mass, defined molecular size and shape for avoiding biological barriers.[77]

2.3.5 Other Cationic Polymers

Includes cationic polyesters, poly(amino ester)s, poly(amidoamine)s, etc.

2.3.5.1 Cationic polyesters

Polyesters have hydrocarbon backbones that contain ester functional groups or linkages. Some of the commonly used polyesters and their copolymers for gene delivery mechanisms include polyhydroxyalkanoate (PHA), polycaprolactone (PCL), polyglycolide or polyglycolic acid (PGA), and polylatic acid (PLA). PHA is biological polyesters produced by microorganisms. PHA have gained major importance due to their structural diversity and close analogy to plastics. Additionally, these are also biodegradable and show good biocompatibility.

Considering polyester copolymers, poly(4-hydroxy-L-proline) ester (PHP) is an water soluble polycationic biodegradable polymeric gene carrier. PHP ester forms a strong complex with DNA by means of electrostatic interaction. Transfection of β-galactosidase gene into mammalian cell using PHP ester-DNA complexes showed the possibility of using PHP ester as a biodegradable gene delivery carrier.[78, 79] Lim *et al.* synthesized biodegradable cationic polymer, poly[alpha-(4-aminobutyl)-L-glycolic acid] (PAGA). Because of degradability and the biocompatibility of the monomer, this polymer does not showed any cytotoxicity. Furthermore, PAGA-DNA complexes exhibited higher transfection efficiency in cells and were reported as useful gene delivery agents.[80]

Li and Huang synthesized and characterized copolymer poly(D,L-lactide-co-4-hydroxy-L-proline) (PLHP) as a polymeric gene carrier for sustained gene delivery. The pDNA-loaded microspheres based on the copolymer were prepared by solvent evaporation emulsion method. The release profile of pDNA from PLHP microspheres showed an initial burst and then slower and continuous release for about 18 days. Gene transfer efficiency of PLHP/pDNA delivery system showed a sustained activity (over a week) when compared with PEI and PLL, suggesting that PLHP is a promising candidate for long-term gene delivery with good biocompatibility and biodegradability.[81]

d'Ayala *et al.* describes the synthesis of nanosized particles consisting of PEI (25 kDa) crosslinked with PCL (50–60 kDa). The nanoparticles obtained were characterized in relation to their physicochemical and biological properties, and results are extremely promising in terms of low cell toxicity and high transfection efficiency.[82] Further, to overcome toxic effects of PEI with a minimal impact on transfection efficiency, non-ionic biocompatible polyesters have been conjugated with PEI. Series of PEI and γ-PGA nanocomposites were prepared and evaluated in terms of cell viability and transfection efficiency by Tripathi *et al.* Results obtained demonstrated the potential of the nanocomposites for gene delivery.[34]

Biodegradable PLA and PLGA microsphere/nanoparticles are advance thriving approaches for *in vivo* nucleotide delivery; due to small particles size, they are easy to inject directly to the targeted organs.[83, 84] Munier *et al.* reported biodegradable cationic nanoparticles (cNP) made of PLA as a promising carrier for *in vivo* DNA delivery and immunization. A versatile approach for the use of pre-formed particles and subsequent adsorption of a model polycation (PEI) was described. Two more polycations, chitosan and poly(2-dimethyl-amino) ethyl methacrylate (pDMAEMA) were also considered to determine the most suitable one for the development of PLA cNP as DNA carriers. Cationic PLA-PEI, PLA-chitosan, and PLA-pDMAEMA nanoparticles were compared for interaction with pDNA and its biological properties of bound DNA. The highest transfection efficiency in cell culture was obtained with PLA-pDMAEMA carriers.[85] Similarly, Oster *et al.* formulated DNA nanocarriers of biodegradable polyesters consisting of amine modified poly(vinyl alcohol) (PVAL) backbones grafted with PLGA. Amine modified high MW biodegradable polyesters combining specific characteristics, such as electrostatic interactions between DNA and facilitating loading of DNA.[86]

2.3.5.2 Poly(amino ester)s

For the maximum and novel use of polyesters in therapeutic applications, biodegradable cross-linked poly(amino ester)s have been developed and used as non-viral vectors for gene delivery. They are biodegradable and non-toxic cationic polymer deliberated in terms of molecular mass, polymer end groups, complex size, and N/P ratio (i.e., ratio of protonatable polymer amine groups to nucleic acid phosphate groups).

Poly(amino ester)s have shown great potential with good transfection efficiencies and low cytotoxicity *in vitro* like PEI[87] and can be divided into two types, linear poly(amino ester)s containing tertiary amines in their backbones and hyperbranched poly(amino ester)s containing tertiary amines in the core and primary amines at the periphery. Poly(amino ester)s of different structures containing primary, secondary, and tertiary amines are desirable, because different types of amines play various roles such as the condensation of DNA and forming the pH-buffering "proton sponge" to facilitate escape of vectors from lysosomes.[88]

Guo *et al.* reported a novel biodegradable poly(ester amine) copolymer prepared from low MW PEI (2 kDa) and poly(L-lactide) copolymers. Prepared copolymers showed a good ability to condense pDNA effectively into nanocomplexes with a small particle size (~150 nm). Results obtained indicate potential application for such poly(ester amine) as a gene delivery system.[89] Further, degradable, hyperbranched poly(ester amine)s based on poloxamer diacrylate and low MW branched PEI were successfully synthesized and evaluated as a gene carrier by Kim *et al.* These poly(ester amine)s showed good DNA binding ability with sizes of complexes below

150 nm, implicating its potential for intracellular delivery. It showed lower cytotoxicity, much higher transfection efficiencies in three cell lines compared with PEI.[90] A few hyperbranched poly(amino ester) synthesized showed transfection efficiency comparable to PEI (25 kDa) for *in vitro* DNA delivery.

Wu *et al.* blended poly(amino ester) with PLGA for DNA vaccination, and microparticles prepared were used for significant lessening of tumor size in mice with the release of pDNA for several years.[91,92] Zugates *et al.* synthesized PAE using a new amine monomer, 2-(pyridyldithio)-ethylamine (PDA). These cationic, degradable polymers containing pyridyldithio functionalities were reacted with mercaptoethylamine (MEA) and the thiol peptide RGDC. These two polymer derivatives displayed strong DNA binding; complexes displayed low cellular toxicity and ability to mediate transfection at levels comparable to PEI in human hepatocellular carcinoma cells. These results suggested PDA based poly(amino ester)s as a modular platform for polymer-mediated gene delivery in the therapeutic delivery of DNA.[93]

2.3.5.3 Poly(amido amine)s

Poly(amino amine)s (PAA) polymers are water-soluble, biodegradable, and biocompatible with lower cytotoxicity profiles. They can be easily synthesized by Michael-type polyaddition of primary or bis-secondary amines to bis(acrylamide)s. These PAA have the advantage to degrade into oligomeric products in aqueous media within days or weeks, depending on their structure. PAA are degradable in aqueous solution by means of hydrolytic cleavage of the amidic bonds in their main chain. These polymers have high potential in biomedical applications and are found to be particularly suitable for use as intracytoplasmic and endosomolytic vectors for the delivery of anticancer drugs, proteins, and nucleic acids.[94,95]

Series of novel bioreducible PAA containing multiple disulfide linkages (SS-PAA) were synthesized and evaluated as non-viral gene vectors by Lin *et al.* The SS-PAA polymers are relatively stable in medium mimicking physiological conditions but are rapidly degraded in the medium mimicking the intracellular reductive environment. The results indicate bioreducible PAA have excellent properties for the development of highly potent and non-toxic polymeric gene carriers.[96] Similarly, series of reducible hyperbranched PAA with high transfection efficiency were designed and synthesized as non-viral gene delivery vectors by Chen *et al.* The density of disulfide linkages in the synthesized hyperbranched polymers was tuned by varying the feed molar ratio of the bisacrylamide monomers. The results demonstrate that disulfide content in PAA controls the MW of the polycation degradation products, ease of polyplexes disassembly, polycation cytotoxicity, and polyplexes transfection activity. Significantly increased transfection activity was achieved with optimized formulations of reducible polyplexes compared with non-reducible controls.[97] Vader *et al.* reported PAA with minor fractions of protonatable fragments in the main chain are promising carriers for delivery of siRNA.[98]

2.4 Factors Influencing Cationic Polymer Mediated Nucleotides Delivery

For a successful *in vivo* gene delivery, a vector competent of performing multiple functions at a suitable timing throughout the delivery process is required. There are several factors like surface charge density, MW of polymer, size of polyplexes, etc., that affect the cationic polymer mediated delivery of nucleotides.[99] For cellular internalization, the negatively charged cell membranes are being facilitated by the cationic polymers due to the positive charge. One way to achieve improved uptake into cells is by increasing the positive surface charge of complexes. This modification is expected to result in higher affinity to negatively charged cell membrane constituents, and thus higher rate of uptake can be achieved.

The disproportionate size of uncondensed pDNA makes it more difficult for the cellular uptake. The charge of the carrier can also influence the size of the polyplexes that plays a significant role in the endosomal uptake, the cytoplasmatic transport, and the migration through the nucleoporecomplexes that mediate the bidirectional transport between cytoplasm and nucleus. Bishop reported the accepted size limit for cellular uptake is around 150 nm.[100, 101] Further, MW of cationic polymers often hinders while delivering the therapeutic nucleotides. It has been seen that the high MW cationic polymers have proficient gene transfection and also induce cytotoxicity, while low MW cationic polymers possibly will reduce cytotoxicity and have also reduced gene transfection. Some of the low

MW polymers have reduced DNA condensation abilities and their marginal transfection properties; thus biodegradable polymers of high MW have been developed with reduced toxicity and high gene delivery mechanisms.[102]

2.5 Biomedical Applications

Gene therapy shows much promise in therapies for various genetic diseases, cancers, viral infection, and cardiovascular disorders. Gene delivery includes both the delivery of pDNA (encoding therapeutic proteins) and RNA interference. DNA delivery provides a corrective action by expressing suppressor proteins to suppress cancer growth or activate apoptosis, while RNA interference delivery produces gene silencing to inhibit cellular protein expression or induce apoptosis.[24] The design and development of nucleic acids based therapeutics for the treatment of diseases arising from genetic abnormalities has made significant progress over the past few years. Cationic polymers are the ultimate vectors that are versatile for the nucleic acid delivery with ease of manipulation. Some important biomedical applications of cationic polymers are discussed underneath.

2.5.1 Tumor Therapy

Cationic polymer-based nanoparticles emerged as a promising nanosize carrier systems to the tumor cell level with a wide range of modification and application possibilities.[19] Nanoparticulates based on cationic polymers like PEI, PLL, and chitosan were blended effectively to deliver nucleic acid as well as anticancer agents to tumor cells. Due to the favorable cellular interaction, internalization, and stronger interaction properties, cationic polymers are considered as promising delivery systems for cancer therapy through both systemic and local administration as compared to polymers with neutral or negative charge. d'Ayala *et al.* reported synthesis of nanosized particles consisting of HMW PEI (25 kDa) crosslinked with PCL (50–60 kDa). Results are extremely promising in terms of low cell toxicity and high transfection efficiency. These biological effects might be related to the peculiar DNA binding to covalently connected polymeric nanoparticles, without the formation of entangled DNA/polymer-soluble aggregates.[82] Further, Cao *et al.* synthesized diblock copolymers of PCL and LPEI and assembled it to biodegradable nano-carriers for co-loading siRNA and small molecular drug in a multifunctional hierarchical nano-assembly, for simultaneously delivering siRNA and drug into the same cancer cells, yielding synergistic effect of RNA interference and chemotherapy in cancer.[103]

2.5.2 siRNA Delivery

Regulation of gene expression using short interfering RNA (siRNA) is a promising strategy for research and treatment of numerous diseases. siRNA is a rapidly emerging class of new therapeutic molecules for the treatment of inherited and acquired diseases. However, poor cellular uptake and instability in physiological conditions limit its therapeutic potential. Hence there is a need to develop a delivery system that can protect and efficiently transport siRNA to the target cells. Cationic delivery vectors form condensed complexes with siRNA that, in turn, provides protection to siRNA against enzymatic degradation and further leads to tissue and cellular targeting. siRNA delivery has also been studied widely together with polymers like PEI, PLL chitosan, and PEG-based polymers.

Nanoparticles derived from polymers, such as chitosan and PEI, have found numerous applications for efficient delivery of siRNA, owing to ease of manipulation, high stability, low cost, and high gene carrying capability.[104] Kim *et al.* have developed and characterized a delivery system for siRNA composed of PEI, PEG, and mannose. PEI forms polyplexes with siRNA, PEG forms a hydrophilic layer outside of the polyplexes for steric stabilization, and mannose serves as a cell binding ligand for macrophages. PEGylation of the PEI reduced toxicity without any adverse reduction in knockdown efficiency relative to PEI alone. Mannosylation of the PEI-PEG could be carried out without any significant reduction in knockdown efficiency relative to PEI alone.[105] Further, chitosan is effective when used as *in vitro* and *in vivo* carrier of siRNA and can mediate gene silencing activity with no toxicity. Through complementary siRNA technique and using chitosan, harmful genes can be "silenced" by delivering it to target cells. Aigner has discussed in detail non-viral strategies, based on different siRNA formulations and various modes of administration, for the delivery of therapeutic siRNAs *in vivo*.[106]

2.5.3 DNA Vaccination

DNA vaccination is one of the most significant biomedical applications of cationic polymers and has attracted many researchers to put more effort.[107] Numerous cationic polymers like PEI, poly(amino ester)s, etc., are successively blended for DNA vaccination with significant enhancement in transfection efficiency. DNA-based vaccine directed to tumor-specific antigens is an attractive strategy in cancer prevention and therapy. In view of the poor immunogenicity of most tumor-associated antigens, studies were carried out to examine the adjuvant effect of PEI on the efficacy of cancer vaccine strategy. Results obtained illustrated the potential use of PEI as an adjuvant in DNA-based cancer vaccination for induction of protective and therapeutic immunity.[108] Oster *et al.* studied cationic microparticles for DNA adsorption formulated by blending PLGA with different cationic agents PEI or CTAB (cetyl-trimethyl-ammonium-bromide). Microparticles with 10% PEI efficiently adsorbed DNA and protected DNA from enzymatic degradation, and, thus, PLGA+PEI microspheres can be used as an adjuvant delivery system for DNA.[109]

2.5.4 Lung and Liver Delivery

Gene therapy aimed at the respiratory epithelium holds therapeutic potential for diseases such as cystic fibrosis and lung cancer. PEI and other cationic polymers have been utilized for gene delivery to the airways. Kleemann *et al.* described a new modification of PEI, in which an oligopeptide related to the protein transduction domain of HIV-1 TAT was covalently coupled to PEI (25 kDa) through a heterobifunctional PEG spacer resulting in a TAT-PEG-PEI conjugate. TAT-PEG-PEI represents a new approach to non-viral gene carriers for lung therapy, comprising protection for pDNA, low toxicity, and significantly enhanced transfection efficiency.[110]

Novel cationic polymer, like dextran-spermine (D-SPM), has been found to mediate gene expression in a wide variety of cell lines and *in vivo* through systemic delivery. Recently, Abdullah *et al.* determined the optimal conditions for gene expression of D-SPM/pDNA in cell lines and in the lungs of BALB/c mice via instillation delivery. *In vitro* studies showed that D-SPM could partially protect pDNA from degradation by nuclease and exhibited optimal gene transfer efficiency. Study also showed that there was a clear trend of increasing size of the complexes as the amount of pDNA was increased.[111]

It was also observed, during *in vitro* administration, polyplexe aggregates are formed, and they are entrapped in pulmonary endothelium. In some cases, amendment of the polyplexes with antibody in order to decrease the zeta potential increases the *in vitro* transfection efficiencies in the lung endothelial cells.[112] Similarly, a target receptor is being set in liver delivery technique to deliver the liver-specific polymeric carriers known as asialoglycoprotein receptors, which is exclusively expressed on hepatocytes and mediates cellular uptake through the specific recognition and endocytosis of ligands. Perales *et al.* reported supervision of targeted polyplexes via tail vein injection into mice resulted in increased uptake of polyplexes into the liver and higher liver-specific expression.[113]

2.5.5 Brain Delivery

Brain delivery can be achieved via the systemic delivery or by localized injection to the brain parenchyma, although some limitations are there in case of systematic delivery. Posadas *et al.* describe efficient delivery of genetic material (both DNA and small interfering RNA) to neuronal tissue using non-viral vectors, including the use of cationic lipids, PEI derivatives, dendrimers, carbon nanotubes, and the combination of carbon-made nanoparticles with dendrimers.[114] Barret *et al.* during their study of PLL-based polyplexes concluded that neuronal transfection efficiencies can be enhanced using PLL polymer.[115]

2.6 Conclusion

Non-viral vectors have shown immense potential for successful gene therapy and delivery of therapeutics nucleotides. Without a doubt, polycations represent the effective carriers for polymer/nucleotide complexes. It represents extremely promising research area where remarkable efforts have been made to apply various different cationic polymers as vectors in gene delivery. Cationic polymers are more advantageous as compared to viral vectors because of their low immune response against the cells and degradation ability, i.e., biodegradability owing to which they easily release the nucleotides for intracellular delivery.

References

1. J. M. Stribley, K. S. Rehman, H. Niu, and G. M. Christman, *Fertil Steril*, **77**, 645 (2002).
2. S. Akhtar, M. D. Hughes, A. Khan, M. Bibby, M. Hussain, Q. Nawaz, *et al.*, *Adv. Drug Deliv. Rev.*, **44**, 3 (2000).
3. S. D. Patil, D. G. Rhodes, and D. J. Burgess, *AAPS J*, **7**, E61 (2005).
4. T. Azzam and A. J. Domb, *Curr. Drug Deliv.*, **1**, 165 (2004).
5. R. G. Crystal, *Science*, **270**, 404 (1995).
6. C. Louise, *Methods Mol. Biol.*, **333**, 201 (2006).
7. F. D. Ledley, *Hum. Gene Ther.*, **6**, 1129 (1995).
8. S. Ferrari, D. M. Geddes, and E. W. Alton, *Adv. Drug Deliv. Rev.*, ***54***, 1373 (2002).
9. D. Luo, and W. M. Saltzman, *Nat. Biotechnol.*, **18**, 33 (2000).
10. K. Kaminski, M. Plonka, J. Ciejka, K. Szczubialka, M. Nowakowska, B. Lorkowska, *et al.*, *J. Med. Chem.*, **54**, 6586 (2011).
11. I. Yudovin-Farber, C. Yanay, T. Azzam, M. Linial, and A. J. Domb, *Bioconjug Chem.*, **16**, 1196 (2005).
12. M. Thanou, B. I. Florea, M. Geldof, H. E. Junginger, and G. Borchard, *Biomaterials*, **23**, 153 (2002).
13. S. Prabha, W. Z. Zhou, J. Panyam, and V. Labhasetwar, *Int. J. Pharm.*, **244**, 105 (2002).
14. J. K. Vasir, and V. Labhasetwar, *Expert Opin. Drug Deliv.*, **3**, 325 (2006).
15. N. Smyth Templeton, *DNA Cell Biol.*, **21**, 857 (2002).
16. S. C. De Smedt, J. Demeester, and W. E. Hennink, *Pharm. Res.*, **17**, 113 (2000).
17. H. Eliyahu, A. Makovitzki, T. Azzam, A. Zlotkin, A. Joseph, D. Gazit, *et al.*, *Gene Ther.*, **12**, 494 (2005).
18. X. B. Xiong, H. Uludag, and A. Lavasanifar, *Biomaterials*, **30**, 242 (2009).
19. E. Bilensoy, *Expert Opin. Drug Deliv.*, **7**, 795 (2010).
20. G. A. Pietersz, C. K. Tang, and V. Apostolopoulos, *Mini. Rev. Med. Chem.*, **6**, 1285 (2006).

21. X. Sun, and N. Zhang, *Mini. Rev. Med. Chem.*, **10**, 108 (2010).
22. A. Akinc, M. Thomas, A. M. Klibanov, and R. Langer, *J. Gene Med.*, **7**, 657 (2005).
23. U. Lungwitz, M. Breunig, T. Blunk, and A. Gopferich, *Eur. J. Pharm. Biopharm.*, **60**, 247 (2005).
24. F. J. Xu, and W. T. Yang, *Prog. Polym. Sci.*, **36**, 1099 (2011).
25. W. Dong, G. H. Jin, S. F. Li, Q. M. Sun, D. Y. Ma, and Z. C. Hua, *Acta. Biochim. Biophys Sin (Shanghai)*, **38**, 780 (2006).
26. T. Chen, Z. Wang, R. Wang, T. Lu, and W. Wang, *J. Drug Target*, **15**, 714 (2007).
27. J. Y. Cherng, W. C. Hung, and H. C. Kao, *Curr. Pharm. Biotechnol.*, **12**, 839 (2011).
28. M. Breunig, U. Lungwitz, R. Liebl, C. Fontanari, J. Klar, A. Kurtz, *et al.*, *J. Gene Med.*, **7**, 1287 (2005).
29. Y. J. Liu, Y. P. Li, J. Li, J. F. Jia, and L. Liu, *Shi Yan Sheng Wu Xue Bao*, **37**, 91 (2004).
30. J. Li, Q. Wang, H. Yu, F. Sheng, W. Ma, and P. Yi, *Sheng Wu Yi Xue Gong Cheng Xue Za Zhi*, **22**, 1210 (2005).
31. R. Goyal, S. K. Tripathi, S. Tyagi, A. Sharma, K. R. Ram, D. K. Chowdhuri, *et al.*, *Nanomedicine*, **8**, 167 (2012).
32. M. L. Forrest, J. T. Koerber, and D. W. Pack, *Bioconjug. Chem.*, **14**, 934 (2003).
33. W. Liang, H. Gong, D. Yin, S. Lu, and Q. Fu, *Chem. Pharm. Bull. (Tokyo)*, **59**, 1094 (2011).
34. S. K. Tripathi, R. Goyal, K. M. Ansari, K. Ravi Ram, Y. Shukla, D. K. Chowdhuri, *et al.*, *Eur. J. Pharm. Biopharm.*, **79**, 473 (2011).
35. J. Dai, S. Zou, Y. Pei, D. Cheng, H. Ai, and X. Shuai, *Biomaterials*, **32**, 1694 (2011).
36. W. C. Tseng, T. Y. Fang, L. Y. Su, and C. H. Tang, *Mol. Pharm.*, **2**, 224 (2005).
37. K. M. Lee, I. S. Kim, Y. B. Lee, S. C. Shin, K. C. Lee, and I. J. Oh, *Arch. Pharm. Res.*, **28**, 722 (2005).
38. K. M. Lee, Y. B. Lee, and I. J. Oh, *J. Nanosci. Nanotechnol.*, **11**, 7078 (2011).
39. J. A. Fortune, T. I. Novobrantseva, and A. M. Klibanov, *J. Drug Deliv.*, **2011**, 204058 (2011).
40. C. Hu, Q. Peng, F. Chen, Z. Zhong, R. Zhuo, *Bioconjug. Chem.*, **21**, 836 (2010).

41. A. Sharma, A. Tandon, J. C. Tovey, R. Gupta, J. D. Robertson, J. A. Fortune, *et al.*, *Nanomedicine*, **7**, 505 (2011).

42. A. Gonsho, K. Irie, H. Susaki, H. Iwasawa, S. Okuno, and T. Sugawara, *Biol. Pharm. Bull.*, **17**, 275 (1994).

43. A. Maruyama, H. Watanabe, A. Ferdous, M. Katoh, T. Ishihara, and T. Akaike, *Bioconjug. Chem.*, **9**, 292 (1998).

44. W. T. Godbey, K. K. Wu, and A. G. Mikos, *J. Control Release*, **60**, 149 (1999).

45. O. Boussif, T. Delair, C. Brua, L. Veron, A. Pavirani, and H. V. J. Kolbe, *Bioconjugate Chemistry*, **10**, 877 (1999).

46. Y. H. Choi, F. Liu, J. S. Kim, Y. K. Choi, J. S. Park, and S. W. Kim, *J. Control Release*, **54**, 39 (1998).

47. M. I. Nounou, K. Emmanouil, S. Chung, T. Pham, Z. Lu, and M. Bikram, *J. Control Release*, **143**, 326 (2010).

48. C. Fu, X. Sun, D. Liu, Z. Chen, Z. Lu, and N. Zhang, *Int. J. Mol. Sci.*, **12**, 1371 (2011).

49. L. L. Farrell, J. Pepin, C. Kucharski, X. Lin, Z. Xu, and H. Uludag, *Eur. J. Pharm. Biopharm.*, **65**, 388 (2007).

50. M. Yamagata, T. Kawano, K. Shiba, T. Mori, Y. Katayama, and T. Niidome, *Bioorg. Med. Chem.*, **15**, 526 (2007).

51. X. Zhang, M. Oulad-Abdelghani, A. N. Zelkin, Y. Wang, Y. Haikel, D. Mainard, *et al.*, *Biomaterials*, **31**, 1699 (2010).

52. S. W. Kim, *Adv. Genet*, **53**PA, 263 (2005).

53. T. Azzam, H. Eliyahu, A. Makovitzki, M. Linial, and A. J. Domb, *J. Control Release*, **96**, 309 (2004).

54. I. Yudovin-Farber, T. Azzam, E. Metzer, A. Taraboulos, and A. J. Domb, *J. Med. Chem.*, **48**, 1414 (2005).

55. E. Khor, and L. Y. Lim, *Biomaterials*, **24**, 2339 (2003).

56. T. Chandy, and C. P. Sharma, *Biomater Artif Cells Artif Organs*, **18**, 1 (1990).

57. K. L. Douglas, C. A. Piccirillo, and M. Tabrizian, *Eur. J. Pharm. Biopharm.*, **68**, 676 (2008).

58. E. A. Klausner, Z. Zhang, R. L. Chapman, R. F. Multack, and M. V. Volin, *Biomaterials*, **31**, 1814 (2010).

59. K. A. Howard, U. L. Rahbek, X. Liu, C. K. Damgaard, S. Z. Glud, M. O. Andersen, *et al.*, *Mol. Ther.*, **14**, 476 (2006).

60. J. Q. Gao, Q. Q. Zhao, T. F. Lv, W. P. Shuai, J. Zhou, G. P. Tang, *et al.*, *Int. J. Pharm.*, **387**, 286 (2010).

61. W. Weecharangsan, P. Opanasopit, T. Ngawhirunpat, T. Rojanarata, and A. Apirakaramwong, *AAPS PharmSciTech*, **7**, 66 (2006).
62. X. Wang, J. Yao, J. P. Zhou, Y. Lu, and W. Wang, *Pharmazie*, **65**, 572 (2010).
63. H. Yu, X. Chen, T. Lu, J. Sun, H. Tian, J. Hu, *et al.*, *Biomacromolecules*, **8**, 1425 (2007).
64. M. L. Forrest, N. Gabrielson, and D. W. Pack, *Biotechnol. Bioeng.*, **89**, 416 (2005).
65. T. G. Park, J. H. Jeong, and S. W. Kim, *Adv. Drug Deliv. Rev.*, **58**, 467 (2006).
66. S. H. Pun, N. C. Bellocq, A. Liu, G. Jensen, T. Machemer, E. Quijano, *et al.*, *Bioconjug. Chem.*, **15**, 831 (2004).
67. C. Yang, H. Li, S. H. Goh, and J. Li, *Biomaterials*, **28**, 3245 (2007).
68. Q. Y. Jiang, L. H. Lai, J. Shen, Q. Q. Wang, F. J. Xu, and G. P. Tang, *Biomaterials*, **32**, 7253 (2011).
69. J. J. Thomas, M. R. Rekha, and C. P. Sharma, *Int. J. Pharm.*, **389**, 195 (2011).
70. J. J. Thomas, M. R. Rekha, and C. P. Sharma, *Colloids Surf B Biointerfaces*, **81**, 195 (2010).
71. W. C. Tseng, C. H. Tang, and T. Y. Fang, *J. Gene Med.*, **6**, 895 (2004).
72. I. Yudovin-Farber, H. Eliyahu, and A. J. Domb, *Cold Spring Harb Protoc*, (2011).
73. H. Hosseinkhani, T. Azzam, Y. Tabata, and A. J. Domb, *Gene Ther.*, **11**, 194 (2004).
74. T. Dutta, N. K. Jain, N. A. McMillan, and H. S. Parekh, *Nanomedicine*, **6**, 25 (2010).
75. M. X. Tang, C. T. Redemann, and F. C. Szoka, Jr., *Bioconjug. Chem.*, **7**, 703 (1996).
76. H. Eliyahu, Y. Barenholz, and A. J. Domb, *Molecules*, **10**, 34 (2005).
77. S. Sonke, *Eur. J. Pharm. Biopharm.*, **71**, 445 (2009).
78. Y. B. Lim, Y. H. Choi, and J. S. Park, *J. Am. Chem. Soc.*, **121**, 5633 (1999).
79. D. Putnam and R. Langer, *Macromolecules*, **32**, 3658 (1999).
80. Y. B. Lim, S. O. Han, H. U. Kong, Y. Lee, J. S. Park, B. Jeong, *et al.*, *Pharm. Res.*, **17**, 811 (2000).
81. Z. Li, and L. Huang, *J. Control Release*, **98**, 437 (2004).
82. G. G. d'Ayala, A. Calarco, M. Malinconico, P. Laurienzo, O. Petillo, A. Torpedine, *et al.*, *J. Biomed. Mater Res. A*, **94**, 619 (2010).

83. M. N. Kumar, S. S. Mohapatra, X. Kong, P. K. Jena, U. Bakowsky, and C. M. Lehr, *J. Nanosci. Nanotechnol.*, **4**, 990 (2004).

84. M. N. Ravi Kumar, U. Bakowsky, and C. M. Lehr, *Biomaterials*, **25**, 1771 (2004).

85. S. Munier, I. Messai, T. Delair, B. Verrier, and Y. Ataman-Onal, *Colloids Surf B Biointerfaces*, **43**, 163 (2005).

86. C. G. Oster, M. Wittmar, U. Bakowsky, and T. Kissel, *J. Control Release*, **111**, 371 (2006).

87. D. M. Lynn. and R. Langer, *J. Am. Chem. Soc.*, **122**, 10761 (2000).

88. Y. Liu, D. Wu, Y. Ma, G. Tang, S. Wang, C. He, *et al.*, *Chem. Commun. (Camb)*, 2630 (2003).

89. Q. F. Guo, T. T. Liu, X. Yan, X. H. Wang, S. Shi, F. Luo, *et al.*, *Int. J. Nanomedicine*, **6**, 1641 (2011).

90. T. H. Kim, S. E. Cook, R. B. Arote, M. H. Cho, J. W. Nah, Y. J. Choi, *et al.*, *Macromol. Biosci.*, **7**, 611 (2007).

91. D. Wu, Y. Liu, X. Jiang, L. Chen, C. He, S. H. Goh, *et al.*, *Biomacromolecules*, **6**, 3166 (2005).

92. J. Wu, M. A. Mutschler, and C. C. Chu, *J. Mater Sci. Mater Med.*, **22**, 469 (2011).

93. G. T. Zugates, D. G. Anderson, S. R. Little, I. E. Lawhorn, and R. Langer, *J. Am. Chem. Soc.*, **128**, 12726 (2006).

94. G. Coue, and J. F. Engbersen, *J. Control Release*, **152**, 90 (2011).

95. P. Ferruti, M. A. Marchisio, and R. Duncan, *Macromol. Rapid Commun.*, **23**, 332 (2002).

96. C. Lin, Z. Zhong, M. C. Lok, X. Jiang, W. E. Hennink, J. Feijen, *et al.*, *Bioconjug. Chem.*, **18**, 138 (2007).

97. J. Chen, C. Wu, and D. Oupicky, *Biomacromolecules*, **10**, 2921 (2009).

98. P. Vader, L. J. van der Aa, J. F. Engbersen, G. Storm, and R. M. Schiffelers, *Pharm. Res.*, **28**, 1013 (2011).

99. J. Y. Cherng, H. Talsma, R. Verrijk, D. J. Crommelin, and W. E. Hennink, *Eur. J. Pharm. Biopharm.*, **47**, 215 (1999).

100. N. E. Bishop, *Rev. Med. Virol.*, **7**, 199 (1997).

101. D. Oupicky, C. Konak, K. Ulbrich, M. A. Wolfert, and L. W. Seymour, *J. Control Release*, **65**, 149 (2000).

102. J. Luten, C. F. van Nostrum, S. C. De Smedt, and W. E. Hennink, *J. Control Release*, **126**, 97 (2008).

103. N. Cao, D. Cheng, S. Zou, H. Ai, J. Gao, and X. Shuai, *Biomaterials*, **32**, 2222 (2011).

104. S. Nimesh, N. Gupta, and R. Chandra, *Nanomedicine (Lond)*, **6**, 729 (2011).

105. N. Kim, D. Jiang, A. M. Jacobi, K. A. Lennox, S. D. Rose, M. A. Behlke, *et al.*, *Int. J. Pharm.*, (2011).

106. A. Aigner, *Curr. Pharm. Des.*, **14**, 3603 (2008).

107. S. R. Little, D. M. Lynn, Q. Ge, D. G. Anderson, S. V. Puram, J. Chen, *et al.*, *Proc. Natl. Acad. Sci. USA*, **101**, 9534 (2004).

108. Y. F. Ma, and Y. W. Yang, *Eur. J. Pharm. Sci.*, **40**, 75 (2010).

109. C. G. Oster, N. Kim, L. Grode, L. Barbu-Tudoran, A. K. Schaper, S. H. Kaufmann, *et al.*, *J. Control Release*, **104**, 359 (2005).

110. E. Kleemann, M. Neu, N. Jekel, L. Fink, T. Schmehl, T. Gessler, *et al.*, *J. Control Release*, **109**, 299 (2005).

111. S. Abdullah, W. Y. Wendy-Yeo, H. Hosseinkhani, M. Hosseinkhani, E. Masrawa, R. Ramasamy, *et al.*, *J. Biomed. Biotechnol.*, **2010**, 284840 (2010).

112. J. A. Hughes, and G. A. Rao, *Expert Opin. Drug Deliv.*, **2**, 145 (2005).

113. J. C. Perales, T. Ferkol, H. Beegen, O. D. Ratnoff, and R. W. Hanson, *Proc. Natl. Acad. Sci. USA*, **91**, 4086 (1994).

114. I. Posadas, F. J. Guerra, and V. Cena, *Nanomedicine (Lond)*, **5**, 1219 (2010).

115. L. B. Barrett, M. Berry, W. B. Ying, M. N. Hodgkin, L. W. Seymour, A. M. Gonzalez, *et al.*, *J. Gene Med.*, **6**, 429 (2004).

Chapter 3

Membrane/Core Nanoparticles for Delivery of Therapeutic Nucleic Acid

Younjee Chung and Leaf Huang*

Division of Molecular Pharmaceutics, UNC Eshelman School of Pharmacy, University of North Carolina at Chapel Hill, Chapel Hill, NC 27599, USA

*leafh@unc.edu

Gene inhibitors, especially small interfering RNA (siRNA), have been widely studied and are considered to have significant importance in therapeutic application for various diseases. However, several limitations related to delivery of those gene inhibitors have delayed advancement in siRNA therapy. Physical and chemical instability and rapid clearance of genes have been major problems in achieving effective and efficient gene delivery. There have been extensive studies to try to overcome these barriers using several different types of delivery systems for both cellular and systemic delivery. The ultimate goal of these studies is to develop an effective gene delivery system, which can improve the stability of nucleic acid with high biocompatibility. This system should have the capability to induce cellular internalization and endosomal escape, followed by a release of nucleic acid into the cytosol. Despite intensive efforts in developing an effective nanocarrier system, efficient delivery of

Nanotechnology for the Delivery of Therapeutic Nucleic Acids
Edited by Dan Peer

ISBN 978-981-4411-04-2 (Hardcover), 978-981-4411-05-9 (eBook)
www.panstanford.com

nucleic acid is still a rate-limiting step for gene therapy. Here, we have developed two types of membrane/core Nanoparticles (NPs) in order to efficiently deliver siRNA and further its use in therapeutic applications. The first LPD (Liposome-Polycation-DNA) formulation showed efficient delivery of siRNA and a potential therapeutic effect for metastatic lung tumor treatment. The second formulation, referred to as LCP (Lipid-Calcium-Phosphate), was developed in order to improve the limited release of cargo of LPD formulation, which was done by replacing the protamine-DNA core with a pH-sensitive calcium phosphate (CaP) core.

3.1 Introduction

Gene therapy has emerged as a promising means to treat various genetic or acquired diseases by altering the genetic make-up of cells. There are three essential types of nucleic acid drugs based on their therapeutic relevance: gene inhibitors, gene vaccines, and gene substitutes.[1] Among these, gene inhibitors, in particular, such as single-stranded antisense oligonucleotide (ODN) and siRNA have been intensively studied and are playing a major role in therapeutic application for many disease treatments.

There are several ways to induce transient disruption of gene expression and further generate a gene silencing effect, which has extensive applications in biomedical research. The first method is to inhibit the translation of a target gene by hybridizing it to the corresponding mRNA using synthetic, small, and single-stranded ODN.[2] Ribozyme, a RNA enzyme, can also be used to cleave single-stranded regions in RNA through esterification or hydrolysis reactions.[3, 4] Double-stranded siRNA shows an improved silencing effect[5] over ODN, with outstanding target specificity and stability in serum. Aside from the improved efficiency on gene silencing, siRNA has other advantages. It does not require a cellular expression system, complex protein purification, or refolding schemes and therefore leads to relatively uncomplicated synthesis. Thus, siRNA has been widely considered as a promising candidate in developing new therapeutic applications for various diseases. Short hairpin RNA (shRNA) is converted to a siRNA duplex by a cellular enzyme, Dicer, and is subsequently incorporated into target mRNA degradation, leading to the silencing of a target gene. Unlike other abundant cellular RNAs, shRNA is transcribed by RNA polymerase

III.[6–10] The structural characteristics of the stems and loops may vary depending on the promoters, but more often shRNA produce efficient target gene silencing.[11] Finally, microRNA interacts partially with mismatched sequences in the 3′-untranslated regions of the target mRNA, and this results in the suppression or degradation of the target mRNA.[12]

Despite the recognized potential of the nucleic-acid-based therapeutics, there are still several limitations to overcome in order to achieve sufficient therapeutic effects. So far, most of the *in vivo* studies for nucleic acid have used high doses of unformulated or "naked" nucleic acid delivered through a hydrodynamic injection.[13–17] When naked nucleic acid is systemically administered, it is highly susceptible to a rapid enzymatic degradation by serum nucleases, resulting in a short half-life ranging from several minutes to about an hour.[18–21] Also, due to its highly charged nature, it does not freely diffuse through the cell membranes, which might be responsible for a low efficiency.[22] Other serious problems for non-targeted nucleic acid are the difficulty in delivering it specifically to target cells and rapid excretion through the urine.[23] A biodistribution study of Cy3-labeled siRNA suggested that it was found primarily in the kidney within 20 minutes after injection and was not detected in any other major organs.[23]

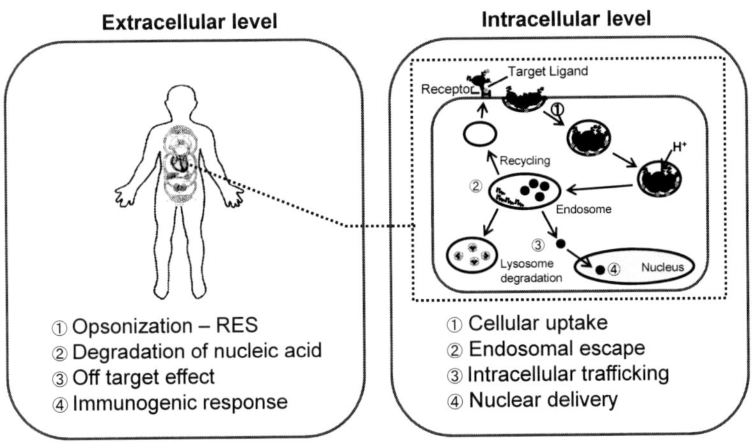

Figure 3.1 Scheme outlining the barriers for gene delivery.

In order to overcome these barriers (Fig. 3.1) and to achieve successful pharmacological effects, an effective and efficient delivery system for both cellular and systemic delivery has to be devised. An ideal gene delivery system should accomplish physical and chemical stability of both the nucleic acid and its vehicle, along with biocompatibility and the capability of transferring exogenous nucleic acid into the target sites with a low risk of immunogenicity and other toxicity.[1] Other considerations that have to be taken into account in developing effective nanocarriers are cellular internalization, endosomal escape, and release of nucleic acid molecules into the cytosol.[24]

Two major carrier systems that show several advantages over the delivery of naked nucleic acid molecules have been developed. These carrier systems are viral and non-viral delivery systems. A viral carrier system using several viruses such as retrovirus, adenovirus, and adeno-associated virus (AAV) has shown efficient transfection properties both *in vitro* and *in vivo*. Viruses are used because of their ability to allow their genomes to penetrate into the target cell's cytoplasm, which may be quite beneficial in overcoming the problem of nucleic acid delivery.[22] However, there is a risk of causing side effects, including immune and toxic responses, random integration of vector DNA into the host chromosome, specific tissue tropism, and potential recombination with wild-type viruses.[1] Therefore, a non-viral delivery system is preferred in gene therapy despite its relatively low efficiency of gene transfer compared with viral carriers. However, recent advances in nanotechnology have significantly increased the gene transfer efficiency of non-viral vectors.

3.2 Challenges in Nanocarrier Systems

Nanotechnology has been continuously applied in the area of non-viral drug delivery in order to develop NPs that can provide improved protection of unstable drugs, such as nucleic acids, for enhanced bioavailability and prolonged therapeutic effects. To study the therapeutic effect and toxicity of NPs, it is essential to understand their pharmacokinetics (PK) profiles and biodistribution. The PK and biodistribution of NPs may be critically affected by their chemical and physical properties such as size, surface charge, and shape. NPs smaller than 5 nm may be rapidly cleared by the kidney.[25,26] Accumulation of larger NPs (greater than 10 nm in size) is often observed in the liver, spleen, and bone marrow because of the limited pore size of the endothelial wall in each tissue.[27–29,58] Since the pore size of liver fenestrate is approximately 100 nm, NPs less than 50 nm were recovered predominantly in the liver (~60%) and resulted in enhanced liver uptake. And the NPs with larger dimensions (~100–200 nm) are known to be cleared by the reticuloendothelial system (RES), i.e., Kupfer cells in the liver.[27] In the spleen, NPs less than 100 nm in diameter showed a minimal splenic uptake, while almost half of the injected dose of NPs with sizes of approximately 400 nm or greater was captured by the spleen. This suggests that the rate of splenic uptake increases as the particle size increases.[30] The reason why the liver, spleen, and bone marrow have to be primarily considered in the PK and biodistribution study of NPs is because these organs largely contain macrophages in their tissues and these macrophages play a key role in clearing NPs and macromolecules from the circulation. This system of macrophages present in each of these tissues is known as a reticuloendothelial system.[31]

The accumulation of NPs in cells of the RES is triggered by the specific serum proteins (opsonin) that are bound on the NPs, a process known as opsonization. The mechanism of opsonization is not clearly understood yet, but major opsonins have been identified. They are fibronectin, immunoglobulins, and complement proteins.[32,33] To minimize opsonization, the surfaces of NPs have to be modified with a hydrophilic polymer, usually polyethylene glycol (PEG). These PEGylated NPs can avoid protein binding by the steric hindrance effect of the polymer chains. The surface density of the PEG chains is a critical factor in determining the degree of reduction in opsonization. A dense (>8 mol%) PEG layer favors the brush conformation and effectively reduces opsonization. PEG densities less than 4 mol% favor the mushroom conformation and do not provide protection against opsonization.[34] Several studies with different types of NPs have shown that the PEGylated particles have a larger AUC than that of non-PEGylated particles or free drugs, indicating that PEGylation leads to a reduction in RES uptake and an increased half-life.[35–40] An alternative way to reduce the RES uptake and prolong the blood circulation of NPs may be to neutralize the surface charge of NPs. Highly negatively charged particles have an increased rate of RES uptake in the liver and a higher rate of clearance compared to those of neutral particles.[41] Since positively charged particles tend to form aggregates with negatively charged blood proteins such as albumin,[42] these particles accumulate mostly in the lung and the liver with a rapid blood clearance when dissociated from the bound serum proteins.

In the development of a new and efficient delivery system for therapeutic nucleic acid, physical and chemical properties of NPs should be carefully considered since the size, shape, and surface chemistry critically determine the PK profile and biodistribution of the particles. One may achieve a prolonged circulation in the blood and a reduction of RES uptake by simply manipulating the size, charge, and surface chemistry of the NPs. However, there are still several problems to be resolved for a successful drug delivery even with improved PK and biodistribution. A challenge then will be how to achieve sufficient intracellular delivery and release of the drug cargo from the particles after internalization. These aspects will be discussed below.

3.3 Current Non-Viral Carrier Systems

There are several different non-viral carriers based on their material composition. It is important to select a suitable material based on the nucleic acid cargo in order to effectively incorporate the cargo and also to enhance the gene silencing effect on the target gene.[2] The material most extensively studied for the non-viral delivery system is liposome. Liposomes are spherical vesicles with an aqueous interior surrounded by a phospholipid bilayer membrane. Liposomes can efficiently carry the encapsulated cargo into the cells by endocytosis and enhance drug delivery. Cationic lipids are usually used to condense negatively charged nucleic acid to form a nanostructure known as a lipoplex.[43] Cationic lipids not only sufficiently encapsulate nucleic acid and protect it from serum degradation but also induce destabilization of the endosomal membrane in order to release the nucleic acid cargo into the cytoplasm.[1,44,45] Since cationic liposomes display a high transfection efficiency and are easy to use, several commercially available liposome formulations have already been developed, e.g., Lipofectamine™ and Oligofectamine™.[46] However, these lipoplexes usually have a short half-life *in vivo* due to their excess positive charges and are also toxic to the host. Therefore, it is essential to consider the balance between stability, efficacy, and potential toxicity based on the electrostatic interactions between nucleic acid and lipids.[47]

Similar to liposomes, cationic polymers may also be used to condense negatively charged nucleic acids via electrostatic interactions. The flexibility of polymers in constructing a broad range of nanostructures such as polyplexes, polymeric micelles, and dendrimers makes polymers suitable to develop as non-viral carriers. Since polymers can be extensively and precisely engineered due to their exclusive physicochemical properties, research efforts have been devoted to improving polymers with greater biocompatibility, stability in biological systems, and intracellular trafficking.[1] Among several different types of carriers, including both synthetic and biological polymers, biological polymers are found to have a higher biocompatibility and a relatively low toxicity. Two naturally formed molecules, chitosan and poly(amine-cyclodextrin), may be representative of biological polymers. Both materials have been widely used for nucleic acid delivery mainly because of their cationic charge, excellent biocompatibility, and biodegradable characteristics. However, chitosan has minimal solubility and a low buffering capacity at physiological pH, resulting in significantly reduced transfection efficiency compared to other cationic polymers such as polyethylenimine (PEI).[1] PEI is the most well-known cationic synthetic polymer composed of a high amount of secondary and tertiary amines. Crowed secondary and tertiary amines in PEI show a good buffering capacity in the weakly acidic pH. In the endosome, PEI promotes proton and chloride anion entry, which induces osmotic swelling of the endosome and the subsequent cargo release into the cytoplasm.[1,48] This is the so-called proton sponge effect. PLGA is one of the first polymers approved by the US Food and Drug Administration for human clinical use and it is both biocompatible and biodegradable.[24] Since PLGA is hydrolyzed over time into natural metabolic products, i.e., glycolic acid and lactic acid, PLGA-based nanocarriers have been considered to have excellent safety profiles.[49–51] Also, several studies on PLGA NPs suggested that the encapsulation efficiency of nucleic acid could be enhanced in addition to the sustained release and improved transfection efficiency by incorporating them with other polymers, such as those containing tertiary amines.[52–54]

In recent years, inorganic nanomaterials have attracted much attention as novel nucleic acid carriers thanks to the technological advances in fabricating inorganic nanomaterials. Furthermore, the surface of NPs can be readily functionalized with other polymers or biomolecules via proper conjugation chemistry, which provides a surpassing edge in developing new nanocarriers for gene therapy. Among several types of metallic NPs, gold NPs (AuNPs) have been widely adopted as a nucleic acid delivery system due to their low toxicity and the ease of modification of surface properties such as charge and hydrophobicity.[55,56] Several functional ligands and biomolecules can be attached onto the surfaces of the AuNPs through strong chemisorption of thiol groups, and furthermore these biomolecules can be released from the particles through the reductive intracellular environment.[57]

Quantum dots (QDs) are semiconductor NPs and have significant advantages over the traditional organic dyes, including enhanced stability and minimal photobleaching.[24] These QDs, therefore, have been utilized in biological applications such as imaging, labeling, detection[58–62] and also have been used as add-ons to the other nanocarrier systems due to their relatively small sizes, typically between 2 to 12 nm.[63,64] However, the surface of QDs can be modified with several functional components, including nucleic acid via conjugation, and such features allow QDs to be useful as nanocarriers for gene delivery.

Magnetic NPs (MNPs) have also been used in nucleic acid delivery because of their unique magnetic properties that allow them to be utilized as a diagnostic tool (primarily in magnetic resonance imaging, MRI) and as targetable nanocarriers for gene therapy. This combination of imaging and targeted therapy is known as "theranostics."[65] Among several other materials, including cobalt or nickel, iron oxide has been widely used for a NP core because of its low toxicity profile. The iron oxide core is usually coated with biocompatible materials (e.g., polysaccharide, synthetic polymer, lipid, protein, etc.) to provide stability in physiological conditions and the opportunity for further modification with macromolecular biotherapeutics. Although the adoption of MNPs in drug delivery system is a relatively new approach, there have been several studies that reported successful delivery of siRNA to cancer cells in mice with an improved plasma pharmacokinetic profile and therapeutic effects.[66,67]

Silica has been investigated as a potential candidate for nanocarriers of nucleic acid because of its natural biocompatibility and low toxicity and its ease of modification and functionalization onto the silanol surface of NPs.[68–70] In particular, mesoporous silica

NPs (MSNs) have been considered as promising nanocarriers for nucleic acid due to their high pore volume and surface area as well as a loading efficiency over 30 wt%, depending on the drug.[71] The high loading capacity of MSNs allows for a tunable range of drug loading in particles in order to minimize the side effects and toxicity induced by the excess use of drug or particles.

3.4 Membrane/Core NPs

To date, only few cases have been reported to have successful treatment of tumors with systemically delivered siRNA.[72–75] The development of an efficient nanocarrier system is the much needed rate-limiting step for RNAi-based therapy. The ideal nanocarrier for siRNA should protect the cargo from nuclease degradation and efficiently deliver the siRNA to the cytoplasm where it mediates gene silencing by inducing degradation of messenger RNAs (mRNAs) containing the complementary sequence.[76–78]

Among various delivery systems for RNAi-based therapy, the most promising non-viral carriers for systemic delivery are the PEGylated NPs with proper surface modification utilizing multiple components. The major components for this system are cationic molecules, such as cationic lipid or polymer, which can interact with negatively charged nucleic acid to form nano-sized particles.[20, 79] Proper surface modification with PEGylation and specific ligands is also important in order to provide prolonged circulation of the NPs in the system as well as targeted delivery with improved cellular bioavailability, resulting in higher accumulation in the target sites. Also, endosomolytic agents, such as fusogenic lipids, can be incorporated to facilitate intracellular release of nucleic acid.[20, 80] Such surface modifications are commonly used to avoid many barriers in systemic delivery and develop an efficient delivery system.

The RES is a major barrier for NPs after systemic administration. When NPs are systemically administered, they are immediately recognized by serum proteins, also known as opsonins, and these NPs are rapidly taken up by macrophages in the RES, located mostly in the liver and the spleen.[81] In the early stages of developing a strategy to avoid the RES, a pre-dosing method was used. A large excess of empty carrier was administered first to saturate the RES, followed by drug-loaded NPs.[82,83] The hepatic uptake was reversibly blocked by this method but not the splenic uptake. Recent strategies focus mostly on the modification of properties of NPs using hydrophilic polymers to sterically shield the hydrophobic surface of NPs and to avoid interaction with opsonins and/or phagocytic cells of the RES.[84] PEG, a hydrophilic and inert polymer, has been widely used as a steric barrier on the surface of NPs, which minimizes protein binding, leading to prolonged circulation with improved pharmacokinetic properties.

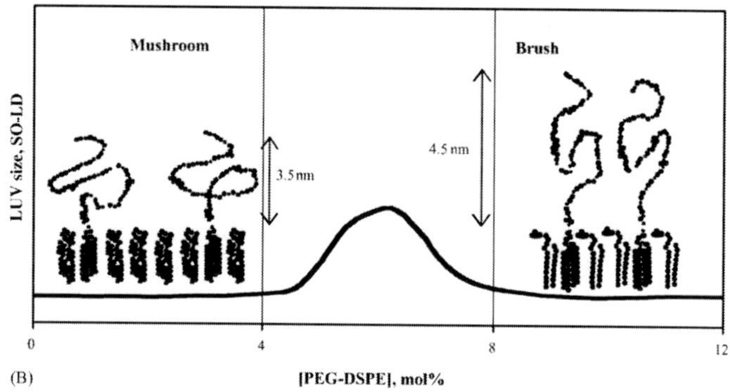

Figure 3.2 PEG conformation on liposome surface.

There are two methods for grafting PEG onto the surface of liposomes. The first one, hydrating a lipid film consisting of both PEG-phospholipids and other lipid components at the same time,[81] allows the PEG to extend from the lipid bilayer to both the inner aqueous core and the outer aqueous phase. The other method, called post-insertion, has been commonly used when PEG-phospholipids are incorporated with pre-formed liposomes[85] by hydrophobic interactions with the acyl chains of liposomal lipids, and permits only the outer surface of liposomes to be modified. In order to avoid opsonization efficiently, it is important that the surface of NPs is sufficiently and stably covered by PEG. The brush model has been known to be the ideal configuration to achieve complete coverage of the surface of the NPs. Theoretically, for a 100 nm liposomal particle modified with DSPE-PEG 2000, there are three possible configurations depending on the mol%: the mushroom conformation with <4 mol% DSPE-PEG 2000; the transition configuration with 4–8 mol%; and the brush mode with >8 mol% modification

(Fig. 3.2).[34] It is also known that having stable brush configuration (>8 mol% DSPE-PEG 2000) is beneficial in order to avoid RES and achieve passive targeting.

Passive targeting has been successfully used as an approach to maximize passive delivery, specifically the EPR effect, which is passive delivery to tumors. However, increasing passive delivery by altering vessel permeability or PEGylation may increase the risk of off-target effects and immune response due to its prolonged systemic circulation.[84] Therefore, an alternative approach for the targeted delivery of NPs has been directed toward the specific receptors or antigens expressed on the surface of tissue/tumor endothelium or tumor cells, which is also referred to as active targeting.[84] There have been extensive studies on tumor targeting using various tumor-specific or organ-specific ligands, such as anisamide or GC4 scFv, conjugated to NPs.[86] The effects of applying active targeting on NPs will be discussed later. Examples of successful nanocarriers are the membrane/core NPs developed in our lab. The following is a summary of these NPs.

3.4.1 LPD

The first developed membrane/core type NP was LPD, which is a self-assembled non-viral nanoparticle (NP) formulation for a systemic delivery of nucleic acid.

3.4.1.1 Formulation of LPD

LPD NPs are composed of nucleic acid (DNA and/or RNA), a polycationic macromolecule, and cationic liposomes. Both LPD and PEGylated LPD NPs have been prepared by combining cationic liposomes and polycation condensed nucleic acid. In general, the formation of NPs with nucleic acid has been done through a self-assembling process mediated mainly by the charge–charge interaction, where cationic carriers bind with anionic nucleic acid. It is known that the *in vivo* transfection efficiency with the cationic lipid, DOTAP, can be improved by incorporating cholesterol into the formulation.[87–89] The liposome containing cholesterol tends to attract less protein and further reduces the liposomal uptake by the RES.[90] Crook *et al.*[91] demonstrated that less than 1% of cells were transfected with DOTAP in the presence of 50% and 100% serum, while DOTAP/cholesterol (1:1 w:w) showed approximately 5% and 8% transfection, respectively. Therefore, small unilamellar cationic liposomes are prepared with DOTAP and cholesterol in 1:1 molar ratio by thin film hydration followed by membrane extrusion for the LPD formulation.

Several research groups have reported that siRNA interacts with cationic carriers with less strength than the plasmid DNA, resulting in the formation of unstable particles and reduced delivery efficiency.[92,93] Self-assembly of macromolecules by charge interaction is largely driven by the loss of the bound counterions and is hence an entropy driven process.[94] Nucleic acids less than 140 nucleotides long are not readily condensed by polycations. Typical siRNA is only 19–22 nucleotides in length and cannot by itself be stably condensed by a polycation, such as protamine. In the case of LPD, an appropriate amount of calf thymus (about 50 kbp) or plasmid DNA (about 6 kbp) was mixed with siRNA to promote the stability of the condensed NPs.[92] The amount of polycationic carriers can be varied according to the nucleic acid cargo of interest. Despite the fact that the plasmid DNA can form a tighter complex with cationic carriers, calf thymus DNA has an advantage as carrier DNA due to its limited amount of immune stimulating CpG motif. Because of the high molecular weight, the LPD particle size can be reduced by 10–30% and the delivery efficiency may be increased by 20–80% when calf thymus DNA is used as carrier DNA.[95] After the mixture of nucleic acid and carrier DNA is prepared (1:1, w:w), a highly positively charged peptide, protamine, may be added to the mixture for further condensation of the negatively charged DNA complex. The delivery efficiency of NPs increases with increasing protamine concentration. However, the net charge of the protamine-DNA complex becomes slightly positive when excess of protamine is added, which eliminates the interaction with the cationic liposome, and results in a reduction of the delivery efficiency. Moreover, the amount of protamine should be adjusted according to the conformation of nucleic acids, as the double stranded siRNA needs more protamines for core condensation than the single stranded AS-ODN. The last step of the self-assembly process is the insertion of PEG-lipid conjugate to the LPD to prevent interaction with serum proteins. The distal end of the PEG chain is tethered with a ligand, such as anisamide, to promote nanoparticle interaction with tumor cells that over-express the sigma receptor. The entire process of self-assembly is schematically shown in Fig. 3.3.

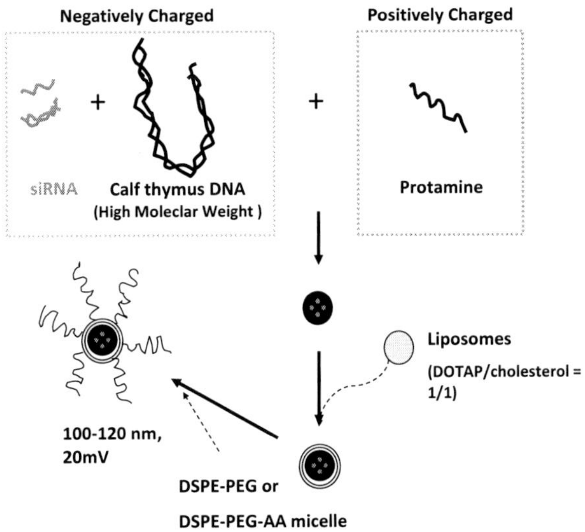

Figure 3.3 The outline for the preparation of LPD NPs.

3.4.1.2 The effect of surface modification of LPD

3.4.1.2.1 *Passive targeting: efficient PEGylation*

In general, the degree of modification with PEG is usually found to be less than 5 mol% if the liposome is to maintain the integrity of the lipid membrane, due to the detergent-like activity of PEG.[85, 96] LPD NPs, however, show a relatively high mol% (~10.6 mol%) of PEG arranged in the brush mode and still remain intact. The key to tolerating the high mol% of PEG arranged in the brush mode may be the presence of two separate lipid bilayers, a unique feature of LPD. The inner layer is supported and stabilized by the charge–charge interaction of the cationic lipids with the negatively charged core. The neutral charge of LPD particles supports the complete shielding provided by the high density of PEG in the brush mode. And these stealthy LPD NPs exhibit a reduced liver uptake in the isolated and perfused liver model (12 ± 5.7% ID at 4 h) and deliver siRNA efficiently into tumor (~30% ID at 4 h).[81] These results indicate that PEGylation plays an important role in preventing serum opsonization and in abolishing the non-specific RES uptake. Since the main loss of the injected NPs by RES comes from the liver and spleen, the low RES uptake of LPD might efficiently help the NPs to reach the tumor with an enhanced permeability and retention (EPR) effect.[97]

3.4.1.2.2 *Active targeting: ligand targeting*

Sigma receptor, a well-known membrane-bound protein, is overexpressed in various types of human tumors, including prostate cancer, melanoma, non-small cell lung carcinoma, and breast tumors, as well as in normal tissues. However, the physiological role of this receptor has not yet been elucidated.[98–104] The neuroleptics[105] and benzamide derivatives,[98] such as anisamide, are known to have a high affinity to the sigma receptor. Huang *et al.*[106] first demonstrated that anisamide, with its small molecular weight, can be successfully used as a ligand to target anticancer drugs to tumors that overexpress sigma receptor. Anisamide is conjugated to the end of PEG lipid and this unique lipid, DSPE-PEG-AA, has been utilized as an active tumor-targeting ligand for LPD NPs (LPD-PEG-AA). *In vitro* studies showed that the delivery efficiency of LPD-PEG-AA for both antisense oligodeoxynucleotide (AS-ODN)[95] and siRNA[95, 107] was 4–7-fold greater than that of LPD-PEG, which has no anisamide ligand but is sterically stabilized by PEGylation, in sigma receptor expressing cells, H1299. Also, when free haloperidol, a well-known high affinity ligand for sigma receptor, is co-treated, it shows partially competitive inhibition in LPD-PEG-AA particles but not in LPD-PEG, confirming that the delivery of AS-ODN or siRNA to H1299 cells by LPD-PEG-AA is mediated by sigma receptor-dependent endocytosis.

The effect of targeting ligand, anisamide, on the enhanced delivery efficiency in H1299 cells has been further supported by the strong silencing effect on target genes, the inhibition of tumor cell growth, and the sensitization of tumor cells to anticancer drugs.[95] The silencing activity of the targeted LPD NPs (LPD-PEG-AA) in a B16F10 lung metastatic tumor mouse model has been found to be significantly higher (70–80% gene silencing) than other formulations, including free siRNA, non-targeted LPD, and control siRNA in the targeted LPD, which appears to be consistent with the *in vitro* observation. This enhanced activity of the LPD-PEG-AA comes mainly from the significantly improved intracellular (cytosolic) delivery, not just from the increased tumor uptake.[107] In another study using a more aggressive lung cancer cell line, NCI-H460, the xenograft tumor mouse model also demonstrated that only the targeted NPs had a significant cytosolic delivery of siRNA. Non-targeted NPs in this study showed less intracellular delivery efficiency, regardless of the comparable amount of accumulation in the tumor (70–80 ID%/g).

This observation has been further confirmed by the z-axis images of each formulation, where a majority of siRNA in the non-targeted NPs formulation remained in the extracellular space but most of the siRNA in the targeted NPs was internalized into the cells.

One of the widely used targeting ligands for gene delivery to cancer is monoclonal antibody. GC4 single-chain variable fragment (scFv), a tumor-targeting human monoclonal antibody,[108] has several advantages as a target ligand, such as its high specificity, strong affinity, profound penetration into the tumor site, and a low antigenicity. Due to these beneficial characteristics of GC4 scFv, it has been used as a targeting ligand in modified LPD NPs (LPH, Liposome-polycation-hyaluronic acid) for the delivery of siRNA and miRNA to a B16F10 lung metastasis mouse model. The GC4 scFv-targeted NPs showed a significant intracellular delivery of siRNA to the tumor tissue and a reduced uptake in the liver and the kidney compared to free siRNA. This indicates that the high delivery efficiency of siRNA or miRNA into the B16F10 lung metastasis mouse is from the tumor-targeting scFv ligand. According to previous studies using targeting ligands, the alteration of intratumoral distribution and increased tumor cell uptake of NPs have been found when tumor-specific ligands, such as anisamide and GC4 scFv, are incorporated into the NPs. Furthermore, the uptake of NPs in tumors may not ultimately be increased when these tumor-specific ligands are not used.[109–113]

3.4.1.3 Therapeutic applications of LPD

3.4.1.3.1 *Survivin siRNA*

Survivin, a 16.5 kDa protein with a single BIR domain, is a member of the inhibitor of apoptosis protein (IAP) family (XIAP, cIAP1, cIAP2, and survivin), which is expressed in human cancer cells.[95] It is a nuclear cytoplasm shuttling protein and is known to inhibit apoptosis and regulate the cell cycle. Overexpression of survivin may lead to an unfavorable progression of disease, development of resistance to chemotherapy as well as radiotherapy, or decreased survival rates in non-small cell lung cancer (NSCLC) patients.[114–116] Therefore, survivin has been intensively utilized as a prognostic marker or potential target for lung cancer treatment. The survivin gene silencing effects of different siRNA formulations were determined in NCI-1299, sigma receptor-expressing cells.

Both mRNA and protein levels of survivin were down-regulated 30% and 70% in LPD-PEG and LPD-PEG-AA NPs, respectively, while the scrambled siRNA showed a reduced effect (20% and not statistically significant). Also, the down-regulation of survivin induces significant apoptosis and remarkable cellular growth inhibition effects in a dose dependent manner.[95]

In a xenograft mouse model using NCI-H460 human lung cancer cells, both non-targeted and anisamide-targeted LPD NPs showed a predominant accumulation in the tumor. However, free FAM-siRNA exhibited little tissue uptake due to the high susceptibility of siRNA to degradation, followed by fast elimination from the systemic circulation. Since there was no difference between non-targeted and targeted LPD NPs in tumor uptake, a targeting ligand is likely to have no influence on the biodistribution of the NPs *in vivo*.[95] Thus, in this case, anisamide served as an internalization, but not targeting, ligand for the NPs.

Based on *in vitro* study and biodistribution study in the mouse model, tumor-targeted PEGylated LPD NPs showed a remarkable gene silencing effect mediated by siRNA with an increased cellular uptake via specific receptor-mediated endocytosis. The fact that the targeted LPD formulation was able to deliver siRNA predominantly to the tumor with a low RES uptake suggests that LPD NPs could be a plausible delivery system for RNAi-based tumor therapy.

3.4.1.3.2 *EGFR siRNA*

The epidermal growth factor receptor (EGFR) is related to many of the negative features of tumors, including decreased apoptosis, increased proliferation, and resistance to chemo as well as radiation therapy[117–120] and is frequently over-expressed in various type of tumors. Several studies have demonstrated that anti-EGFR therapy via tyrosine kinase inhibitors and monoclonal antibodies has a great therapeutic effect in cancer patients.[117–120] As RNAi has emerged as a promising therapy for cancer, silencing EGFR via RNAi has been considered as an alternative approach to anti-EGFR therapy.[121–126] EGFR siRNA delivered by the targeted LPD showed excellent silencing activity in NCI-H460 tumor, whereas the same siRNA delivered by non-targeted NPs exhibited only partial effect. Both free siRNA and the control siRNA in targeted NPs revealed no effect. Silencing EGFR in the tumor was associated with tumor cell apoptosis and tumor growth inhibition. However, maximal anticancer activity was only realized when siRNA therapy was combined with cisplatin chemotherapy.[92] The partial activity of EGFR siRNA was related to the anti-apoptosis activity of DOTAP, a component of the LPD formulation.[92]

3.4.1.3.3 Combination treatment: MDM2, c-myc, and vascular endothelial growth factor

A combination treatment with several different siRNAs might be synergistic in anti-proliferation of the tumor. Based on the study by Song *et al.*,[127] enhanced anti-proliferation on the murine B16 melanoma cells has been observed by the simultaneous silencing of three oncogenes, MDM2, c-myc, and vascular endothelial growth factor (VEGF). MDM2 is an inactivator of the tumor suppressor p53 gene. c-myc acts as an activated transcription factor, which promotes cell proliferation. Lastly, VEGF is an angiogenesis factor that is related to metastasis.[128–130] The combined siRNAs were encapsulated in the anisamide-targeted LPD formulations. Two consecutive IV injections at 0.45 mg/kg, a relatively low dose, showed a significant inhibition effect in lung metastases-bearing mice. Only the siRNAs in targeted NPs showed an effective synergistic effect of three siRNAs, which was consistent with the Western blotting analysis data. It also significantly reduced the number of metastasis nodules in the lung, while other formulations showed little therapeutic effect. This suggested that the silencing activity was highly dependent on the formulation and siRNA sequence. Also, the survival rate of mice treated with siRNAs in the targeted NPs reached the highest point (90%) among all the treatment groups on the 23rd day. Both the non-targeted and the targeted NPs significantly improved the survival of animals compared to that of the free siRNA ($p < 0.05$).[131] At the therapeutic dose (0.45 mg/kg), the targeted NP did not significantly elevate the level of any of cytokines, including interleukin-6 (IL-6), IL-12, tumor necrosis factor, and interferon-α, even after two consecutive injections. The targeted NP was safe at the therapeutic dose.

3.4.1.4 Modified LPD formulations

3.4.1.4.1 *LPD with guanidinium-containing cationic lipid*

Drug resistance in cancer cells is usually caused by the overexpression of drug efflux transporter proteins, such as P-glycoprotein

(P-gp).[132] Since drug resistance is a key obstacle in the successfulness of chemotherapy, the main efforts in cancer research have been devoted to find a proper strategy to tackle the P-gp function by either down-regulating its expression or inhibiting its function. One approach to overcoming multi-drug resistance (MDR) is to utilize NPs to entrap the free drug and deliver it into the cytoplasm or lysosome so that the drug cannot be effluxed out of the cancer cell in a process mediated by P-gp. In order to overcome MDR, N,N-distearyl-N-methyl-N-2-(N′-arginyl) aminoethyl ammonium chloride (DSAA) has been used instead of DOTAP in LPD NPs. DSAA is a cationic lipid, with both guanidinium and lysine residue in its head group, that can inhibit P-gp activity and ultimately increase drug accumulation. This mechanism has been confirmed by incubating doxorubicin with either DSAA or other cationic lipids with no guanidinium group. The lipids without guanidinium moiety only slightly increase doxorubicin uptake, while DSAA significantly enhances doxorubicin accumulation. The guanidinium group of DSAA induces reactive oxygen species (ROS) by donating an electron and also by down-regulating mitogen-activated protein kinase (MAPK) signaling pathways, which results in the down-regulation of the MDR transporter expression.[133]

In addition to its intrinsic therapeutic effect, DSAA can also be utilized as a component of LPD NPs to co-deliver therapeutic siRNA and small molecule drug. One may expect a synergistic therapeutic effect from both drugs. Targeted LPD with DSAA, loaded with both doxorubicin and VEGF siRNA, increases tumor uptake and also suppresses tumor growth by overcoming P-gp mediated drug resistance. However, non-targeted NPs containing DSAA or targeted NPs with DOTAP have shown a reduced therapeutic effect, compared to that of the targeted LPD containing DSAA with both doxorubicin and VEGF siRNA. This suggests that the efficiency of the therapeutic effects of the particles depends on both the lipid and the ligand.[134] Although DSAA showed a preeminent potential both as a formulation component of a delivery carrier and as a therapeutic agent that inhibited MDR expression, it still showed a limitation in its use. The competent interaction of doxorubicin and cationic lipid with therapeutic DNA caused a poor entrapment efficiency of doxorubicin (only ~10%) in NPs. Low entrapment of doxorubicin in NPs could be resolved by replacing the cationic lipid with the anionic lipid, which allowed DNA to interact only with protamine and doxorubicin. This is the LPD-II (anionic liposome-polycation-DNA) formulation discussed below.

3.4.1.4.2 *LPD-II*

In order to increase the entrapment efficiency of doxorubicin in LPD NPs, LPD-II containing anionic liposome instead of cationic liposome has been developed.[134] LPD-II encapsulated approximately 90% of the total doxorubicin, which was significantly enhanced from that of LPD with guanidinium containing cationic lipid (~10% entrapment). LPD-II could more efficiently deliver both doxorubicin and siRNA than could non-targeted NPs or free drug. The higher entrapment rate of doxorubicin in targeted NPs led to a remarkable increase in uptake in the tumor and resulted in an enhanced therapeutic effect by avoiding P-gp mediated drug efflux.[134] Simultaneous inhibition of both MDR and survival signaling in cancer cells was achieved by co-delivery of doxorubicin and a therapeutic siRNA using targeted LPD-II NPs. Since c-Myc positively controls MDR expression,[135–137] down-regulation of MDR could be achieved by silencing c-Myc. Besides c-Myc itself is known as an oncogene; silencing of c-Myc may have an additional direct anticancer effect. Co-delivery of both doxorubicin and c-Myc siRNA in the same targeted LPD-II formulation significantly inhibited tumor growth and enhanced tumor apoptosis in a synergistic manner. Other formulations containing only c-Myc siRNA or only doxorubicin had shown little therapeutic effect. c-Myc together with doxorubicin in the non-targeted NPs had shown only a partial inhibition of tumor growth, indicating that only the targeted NPs with both doxorubicin and c-Myc siRNA could overcome the P-gp mediated drug resistance.[134]

3.4.1.4.3 *LPH*

The proper selection of biopolymer greatly influences the particle condensation, size, and stability of the assembled NPs.[138] The LPD NPs used in previous studies incorporate a high molecular weight polyanion carrier, calf thymus DNA, and successfully increase the delivery efficiency (20–80%). The particle size is also reduced (10–30%) when compared to the formulation without the carrier DNA.[139] Due to its low molecular weight, siRNA forms a loose and unstable complex with a cationic carrier, such as a cationic polymer. This problem can be avoided by incorporating a high molecular weight polyanion with siRNA.[92, 93] Despite its great advantage in improving the delivery efficiency of NPs, there are several concerns with calf thymus DNA, such as the immunogenic response.[140] The FDA is unlikely to approve the use of DNA from an animal source for clinical trials.

In order to minimize the immunotoxicity of the particles, an alternative biocompatible anionic biopolymer, hyaluronic acid (HA), may be used to replace the high molecular weight carrier DNA. HA is a polyanionic polysaccharide, which is one of the major extracellular components of connective tissues.[141,142] Recently, HA has been found to bind to tumor cells that overexpress CD44 when being administered as an anticancer therapy.[143–146] Aside from the beneficial targeting properties, the most pronounced effect of replacing carrier DNA with HA is the relatively low toxicity and immunogenicity, owing to the absence of immune stimulatory CpG motifs. HA has already been approved by the FDA for injection, which is a great benefit in drug development.

The size and zeta potential of the LPH NPs were found to be comparable to those of previous formulation, LPD, regardless of the targeting moiety. This indicates that the replacement of calf-thymus DNA with HA did not affect the physicochemical properties of both formulations. The targeting ligand, anisamide, was found to enhance the siRNA delivery efficiency as well as the gene silencing effect of the LPH NPs in B16F10 cells and in the B16F10 lung metastasis mouse model. Also, LPH NPs did not produce any significant amounts of IL6 and IL12 proinflammatory cytokines over a wide dose range of 0.15–1.2 mg/kg. In contrast, the LPD formulation containing calf thymus DNA significantly increases both cytokines at doses higher than 0.45 mg/kg, limiting the therapeutic dose range. Thus, the LPH formulations can greatly improve the therapeutic window without immunotoxicity while having the same gene silencing effect, when compared to the targeted LPD NPs. The results suggest that LPH might have a greater potential than LPD for clinical use.[139]

3.4.2 LCP

As described above, the LPD formulation has shown efficient delivery of siRNA and a potential therapeutic effect for metastatic lung tumor treatment. However, several possible improvements have been suggested, since the efficiency of release of siRNA from the endosome to the cytosol is limited due to the high affinity of protamine and nucleic acid in the LPD formulation. Therefore, a new formulation with pH-sensitive CaP core replacing the protamine-DNA core in the LPD formulation has been developed to improve the efficiency of cargo release (Fig. 3.4).[147] CaP is considered to be one of the most plausible biomaterials for gene delivery systems with several significant advantages. CaP is known to have a high binding affinity to DNA and other nucleic acids and has been used as a non-toxic transfection reagent for many years. It is biodegradable, biocompatible and dissolves in relatively weak acids.[148] At a low pH condition, such as often found in the endosome, the dissolution of CaP is accelerated. This allows the cargo DNA to be released and escape from the endosome into the cytoplasm, and from there it is ultimately transported to the nucleus for transcription.[149] In addition to these advantages, CaP induces no damage to DNA, no oxidative stress, and no increase in the mutation frequency and thus has been regarded as one of the safest biomaterials so far.[150] There are two types of LCP NPs, LCP-I[147] and LCP-II.[151] Only LCP-II will be discussed here and will be referred to as LCP NPs.

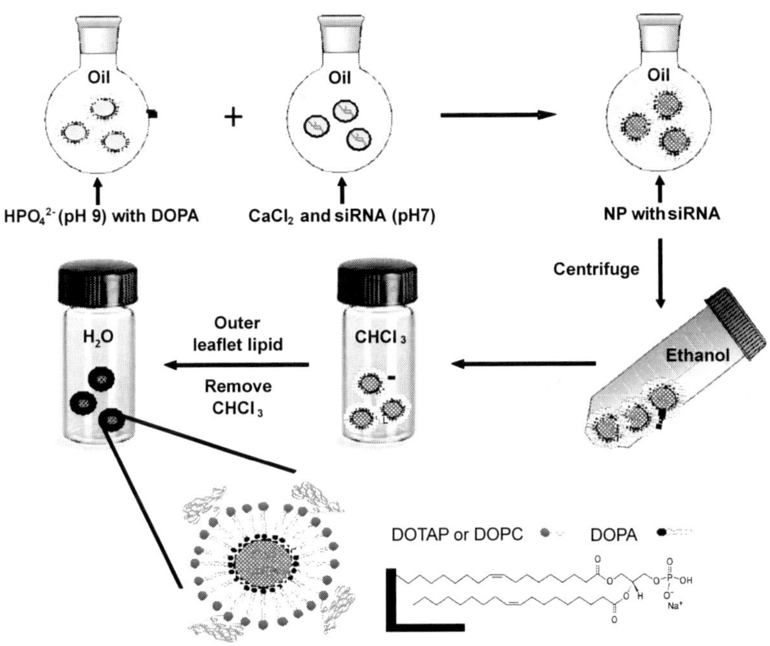

Figure 3.4 The outline for the preparation of LCP-II NPs.

3.4.2.1 Physicochemical characteristic of LCP

An amphiphilic phospholipid, DOPA, plays an important role in the preparation of LCP NPs. Because of the presence of a phosphate group, it is proposed that DOPA participates in the CaP precipitate formation at the interface.[152] The resulting DOPA-coated CaP core is hydrophobic and soluble in organic solvent. This insolubility of the NPs allows a convenient washing of the cores in ethanol to remove excess surfactants such as free DOPA. The size of the final LCP formulation is approximately 40 nm, much smaller than that of LPD (120–150 nm).[80] Since the CaP cores have a hollow structure, it is possible to entrap partially or completely water soluble drugs for a targeted delivery.[151] The hydrophobic core, coated with DOPA, provides a wide range of selection for the outer leaflet lipid in the bilayer surface. Both neutral, e.g., DOPC, and cationic lipid, e.g., DOTAP, may be used as the outer leaflet lipid stabilized by cholesterol.

By comparing the zeta potentials of both pure liposomes and LCP particles using different lipids for the outer leaflet, it was determined that the surface of the NPs is determined mainly by the outer leaflet lipid, not by the inner leaflet lipid, DOPA. The salient features of the LCP particles include the CaP core, in which nucleic acid cargo is encapsulated, and an asymmetrical lipid bilayer coat. Like LPD, the surface lipid bilayer is a supported bilayer that allows the insertion of a large amount of DSPE-PEG with or without a tethered targeting ligand. Two to three fold higher PEG density was required to enhance the tumor uptake of LCP than the corresponding LPD. This is due to the high surface curvature of the LCP compared with that of the LPD.

3.4.2.2 Potential therapeutic effect of LCP

The rapid dissolution of CaP at acidic pH conditions is the main reason for the replacement of protamine-DNA core with CaP to facilitate an enhanced release efficiency of the target cargo. From the study of the siRNA delivery efficiency of LCP in H460 cells *in vitro*, IC_{50} of LCP (5 nM) was 40 times lower than that of LPD (200 nM), indicating that LCP was more effective in siRNA delivery than the LPD with no CaP core. The IC_{50} for LCP was 5 nM regardless of the outer leaflet lipid, DOPC or DOTAP, which suggests that the endosome destabilization effect depended on the

CaP core and not the outer leaflet lipid. Based on this observation, it could be concluded that using cationic lipid for binding with the negatively charged endosomal membrane lipid may not be required to release the cargo from LCP NPs. Unlike in the *in vitro* study, only the LCP NPs containing DOTAP showed significant down-regulation of the target gene (approximately 50%) when 0.6 mg/kg of siRNA was injected to the H460 xenograft mouse model. LCP with DOPC, a neutral lipid, as the outer leaflet lipid showed no silencing effect even with the siRNA dose of 0.12 mg/kg. Such inconsistent results may be better understood when the amount of internalized LCP particles in the endosome is considered.

There are two demonstrated mechanisms for endosome destabilization and cargo escape. The first is that the cationic polymers containing secondary or tertiary amines may increase the osmotic pressure inside the endosome by the buffering effect of the crowded amines, also known as the proton sponge effect.[48] The other mechanism is related to the destabilization of both the endosomal membrane and the cationic vector by the formation of ion pairs between the positively charged groups of either cationic polymer or lipid and the negatively charged groups of the endosomal membrane.[153,154] Ion-pair formation significantly reduces the surface hydration of a membrane promoting non-bilayer structures. If there is sufficient amount of internalized LCP in a certain endosome at one time, similar to the *in vitro* condition, the release of nucleic acid will be mostly mediated by the elevated osmotic pressure caused by CaP dissolution in the acidic endosome.[147] However, when there is insufficient amount of NP in the endosome to increase the osmotic pressure, just as the *in vivo* situation, cationic lipid may play an important role in releasing the nucleic acid cargo.

Since LCP has several advantages over LPD, including improved cargo release activity, it has been used as a delivery carrier for combination siRNA therapy for potential treatment of melanoma metastasis in the lung. LCP is a perfect vector to deliver multiple siRNAs for simultaneous obstruction of several signaling pathways. The combination of three siRNAs targeting MDM2, c-Myc, and VEGF has been successfully delivered to the tumor by the targeted LCP, which resulted in a significant reduction of B16F10 metastatic burden in the lung. Targeted LCP with control siRNA failed to show any therapeutic effect. The mean survival period of the animals in the group treated with therapeutic siRNA formulated in the targeted LCP showed significant improvement over those in the untreated group. The anti-metastasis activity of the combined siRNA was highly dependent on the LCP formulation and the sequence of the siRNA.[155]

3.5 Conclusion

There have been extensive studies on gene therapy, especially RNAi, as a promising therapeutic modality in cancer and other diseases. The development of a safe and effective carrier system is the key element for successful gene therapy. An ideal carrier system has to show the following features: high encapsulation efficiency, stability in the circulation, specificity to the target, fine control of loading and release of the gene, prolonged gene expression, and low immunotoxicity. Among many other approaches in developing non-viral carriers, the cationic lipid membrane/core NPs have shown special promise because of their ability to form stable particles with different negatively charged nucleic acids and the relative ease of fabricating both the core and the surface functionalized with a targeting moiety. However, both the ability to overcome cellular barriers (e.g., avoiding the RES uptake, cargo release from the endosome, and intracellular trafficking of the cargo DNA to the nucleus) and a relatively low transfection efficiency compared to the viral vectors still remain as major challenges for future development. Researchers may eventually be able to overcome these challenges by using rapidly advancing nanotechnology to maximize the clinical significance of a new platform of carrier systems.

Acknowledgment

The original work in this laboratory was supported by NIH grants CA129835, CA129421, CA149363, CA151652, and CA151455. We thank Bethany DiPrete for editing the manuscript.

References

1. Pathak, A., S. Patnaik, and K.C. Gupta, Recent trends in non-viral vector-mediated gene delivery. *Biotechnol J*, 2009. **4**(11): pp. 1559–1572.

2. Stahel, R.A., and U. Zangemeister-Wittke, Antisense oligonucleotides for cancer therapy — an overview. *Lung Cancer*, 2003. **41** Suppl. 1: pp. S81–88.
3. Emilsson, G.M., *et al.*, Ribozyme speed limits. *RNA*, 2003. **9**(8): pp. 907–918.
4. Doudna, J.A., and T.R. Cech, The chemical repertoire of natural ribozymes. *Nature*, 2002. **418**(6894): pp. 222–228.
5. Bertrand, J.R., *et al.*, Comparison of antisense oligonucleotides and siRNAs in cell culture and *in vivo*. *Biochem Biophys Res Commun*, 2002. **296**(4): pp. 1000–1004.
6. Lee, N.S., *et al.*, Expression of small interfering RNAs targeted against HIV-1 rev transcripts in human cells. *Nat Biotechnol*, 2002. **20**(5): pp. 500–505.
7. Miyagishi, M., and K. Taira, U6 promoter-driven siRNAs with four uridine 3′ overhangs efficiently suppress targeted gene expression in mammalian cells. *Nat Biotechnol*, 2002. **20**(5): pp. 497–500.
8. Brummelkamp, T.R., R. Bernards, and R. Agami, A system for stable expression of short interfering RNAs in mammalian cells. *Science*, 2002. **296**(5567): pp. 550–553.
9. Paul, C.P., *et al.*, Effective expression of small interfering RNA in human cells. *Nat Biotechnol*, 2002. **20**(5): pp. 505–508.
10. Yu, J.Y., DeRuiter S.L. , and Turner D.L., RNA interference by expression of short-interfering RNAs and hairpin RNAs in mammalian cells. *Proc Natl Acad Sci U S A*, 2002. **99**(9): pp. 6047–6052.
11. Rossi, J.J., Expression strategies for short hairpin RNA interference triggers. *Hum Gene Ther*, 2008. **19**(4): pp. 313–317.
12. Sall, A., *et al.*, MicroRNAs-based therapeutic strategy for virally induced diseases. *Curr Drug Discov Technol*, 2008. **5**(1): pp. 49–58.
13. Xu, J., *et al.*, Effects of small interference RNA against PTP1B and TCPTP on insulin signaling pathway in mouse liver: evidence for non-synergetic cooperation. *Cell Biol Int*, 2007. **31**(1): pp. 88–91.
14. Behlke, M.A., Progress towards *in vivo* use of siRNAs. *Mol Ther*, 2006. **13**(4): pp. 644–670.
15. Li, X., *et al.*, Alleviation of ischemia-reperfusion injury in rat liver transplantation by induction of small interference RNA targeting Fas. *Langenbecks Arch Surg*, 2007. **392**(3): pp. 345–351.
16. Zender, L., and S. Kubicka, Suppression of apoptosis in the liver by systemic and local delivery of small-interfering RNAs. *Methods Mol Biol*, 2007. **361**: pp. 217–226.

17. Wolff, J.A., and V. Budker, The mechanism of naked DNA uptake and expression. *Adv Genet*, 2005. **54**: pp. 3–20.

18. Layzer, J.M., *et al.*, *In vivo* activity of nuclease-resistant siRNAs. *RNA*, 2004. **10**(5): pp. 766–771.

19. Morrissey, D.V., *et al.*, Activity of stabilized short interfering RNA in a mouse model of hepatitis B virus replication. *Hepatology*, 2005. **41**(6): pp. 1349–1356.

20. Zimmermann, T.S., *et al.*, RNAi-mediated gene silencing in non-human primates. *Nature*, 2006. **441**(7089): pp. 111–114.

21. Dykxhoorn, D.M., D. Palliser, and J. Lieberman, The silent treatment: siRNAs as small molecule drugs. *Gene Ther*, 2006. **13**(6): pp. 541–552.

22. Reischl, D., and A. Zimmer, Drug delivery of siRNA therapeutics: potentials and limits of nanosystems. *Nanomed Nanotechnol Biol Med*, 2009. **5**(1): pp. 8–20.

23. Santel, A., *et al.*, A novel siRNA-lipoplex technology for RNA interference in the mouse vascular endothelium. *Gene Ther*, 2006. **13**(16): pp. 1222–1234.

24. Tan, S.J., *et al.*, Engineering nanocarriers for siRNA delivery. *Small*, 2011. **7**(7): pp. 841–856.

25. Li, S.D., and L. Huang, Pharmacokinetics and biodistribution of nanoparticles. *Mol Pharm*, 2008. **5**(4): p. 496-504.

26. Choi, H.S., *et al.*, Renal clearance of quantum dots. *Nature Biotechnology*, 2007. **25**(10): pp. 1165–1170.

27. Petros, R.A., and J.M. DeSimone, Strategies in the design of nanoparticles for therapeutic applications. *Nat Rev Drug Discov*, 2010. **9**(8): pp. 615–627.

28. Moghimi, S.M., *et al.*, An investigation of the filtration capacity and the fate of large filtered sterically-stabilized microspheres in rat spleen. *Biochim Biophys Acta*, 1993. **1157**(3): pp. 233–240.

29. Porter, C.J., *et al.*, The polyoxyethylene/polyoxypropylene block copolymer poloxamer-407 selectively redirects intravenously injected microspheres to sinusoidal endothelial cells of rabbit bone marrow. *FEBS Lett*, 1992. **305**(1): pp. 62–66.

30. Liu, D., A. Mori, and L. Huang, Role of liposome size and RES blockade in controlling biodistribution and tumor uptake of GM1-containing liposomes. *Biochim Biophys Acta*, 1992. **1104**(1): pp. 95–101.

31. Moghimi, S.M., A.C. Hunter, and J.C. Murray, Long-circulating and target-specific nanoparticles: theory to practice. *Pharmacol Rev*, 2001. **53**(2): pp. 283–318.

32. Moghimi, S.M., *et al.*, Methylation of the phosphate oxygen moiety of phospholipid-methoxy(polyethylene glycol) conjugate prevents PEGylated liposome-mediated complement activation and anaphylatoxin production. *FASEB J*, 2006. **20**(14): pp. 2591–2593.

33. Hamad, I., *et al.*, Complement activation by PEGylated single-walled carbon nanotubes is independent of C1q and alternative pathway turnover. *Mol Immunol*, 2008. **45**(14): pp. 3797–3803.

34. Li, S.D., and L. Huang, Stealth nanoparticles: high density but sheddable PEG is a key for tumor targeting. *J Control Release*, 2010. **145**(3): pp. 178–181.

35. Wang, X., *et al.*, HFT-T, a targeting nanoparticle, enhances specific delivery of paclitaxel to folate receptor-positive tumors. *ACS Nano*, 2009. **3**(10): pp. 3165–3174.

36. Torchilin, V.P., Recent advances with liposomes as pharmaceutical carriers. *Nat Rev Drug Discov*, 2005. **4**(2): pp. 145–160.

37. Berry, G., *et al.*, The use of cardiac biopsy to demonstrate reduced cardiotoxicity in AIDS Kaposi's sarcoma patients treated with pegylated liposomal doxorubicin. *Ann Oncol*, 1998. **9**(7): pp. 711–716.

38. Ewer, M.S., *et al.*, Cardiac safety of liposomal anthracyclines. *Semin Oncol*, 2004. **31**(6 Suppl. 13): pp. 161–181.

39. Sharma, G., *et al.*, Liposomes as targeted drug delivery systems in the treatment of breast cancer. *J Drug Target*, 2006. **14**(5): pp. 301–310.

40. Duncan, R., Polymer conjugates as anticancer nanomedicines. *Nat Rev Cancer*, 2006. **6**(9): pp. 688–701.

41. Levchenko, T.S., *et al.*, Liposome clearance in mice: the effect of a separate and combined presence of surface charge and polymer coating. *Int J Pharm*, 2002. **240**(1–2): pp. 95–102.

42. Zhang, J.S., F. Liu, and L. Huang, Implications of pharmacokinetic behavior of lipoplex for its inflammatory toxicity. *Adv Drug Deliv Rev*, 2005. **57**(5): pp. 689–698.

43. de Fougerolles, A., *et al.*, Interfering with disease: a progress report on siRNA-based therapeutics. *Nat Rev Drug Discov*, 2007. **6**(6): pp. 443–453.

44. Spagnou, S., A.D. Miller, and M. Keller, Lipidic carriers of siRNA: differences in the formulation, cellular uptake, and delivery with plasmid DNA. *Biochemistry*, 2004. **43**(42): pp. 13348–13356.

45. Peer, D., *et al.*, Systemic leukocyte-directed siRNA delivery revealing cyclin D1 as an anti-inflammatory target. *Science*, 2008. **319**(5863): pp. 627–630.

46. Kim, W.J., and S.W. Kim, Efficient siRNA delivery with non-viral polymeric vehicles. *Pharm Res*, 2009. **26**(3): pp. 657–666.
47. Elouahabi, A., and J.M. Ruysschaert, Formation and intracellular trafficking of lipoplexes and polyplexes. *Mol Ther*, 2005. **11**(3): pp. 336–347.
48. Boussif, O., *et al.*, A versatile vector for gene and oligonucleotide transfer into cells in culture and *in vivo*: polyethylenimine. *Proc Natl Acad Sci USA*, 1995. **92**(16): pp. 7297–7301.
49. Panyam, J., and V. Labhasetwar, Biodegradable nanoparticles for drug and gene delivery to cells and tissue. *Adv Drug Deliv Rev*, 2003. **55**(3): pp. 329–347.
50. Shive, M.S., and J.M. Anderson, Biodegradation and biocompatibility of PLA and PLGA microspheres. *Adv Drug Deliv Rev*, 1997. **28**(1): pp. 5–24.
51. Faisant, N., J. Siepmann, and J.P. Benoit, PLGA-based microparticles: elucidation of mechanisms and a new, simple mathematical model quantifying drug release. *Eur J Pharm Sci*, 2002. **15**(4): pp. 355–366.
52. Patil, Y., and J. Panyam, Polymeric nanoparticles for siRNA delivery and gene silencing. *Int J Pharm*, 2009. **367**(1–2): pp. 195–203.
53. Pille, J.Y., *et al.*, Intravenous delivery of anti-RhoA small interfering RNA loaded in nanoparticles of chitosan in mice: safety and efficacy in xenografted aggressive breast cancer. *Hum Gene Ther*, 2006. **17**(10): pp. 1019–1026.
54. de Martimprey, H., *et al.*, siRNA nanoformulation against the ret/PTC1 junction oncogene is efficient in an *in vivo* model of papillary thyroid carcinoma. *Nucleic Acids Res*, 2008. **36**(1): p. e2.
55. Daniel, M.C., and D. Astruc, Gold nanoparticles: assembly, supramolecular chemistry, quantum-size-related properties, and applications toward biology, catalysis, and nanotechnology. *Chem Rev*, 2004. **104**(1): pp. 293–346.
56. Ghosh, P., *et al.*, Gold nanoparticles in delivery applications. *Adv Drug Deliv Rev*, 2008. **60**(11): pp. 1307–1315.
57. Giljohann, D.A., *et al.*, Gene regulation with polyvalent siRNA-nanoparticle conjugates. *J Am Chem Soc*, 2009. **131**(6): pp. 2072–2073.
58. Alivisatos, P., The use of nanocrystals in biological detection. *Nat Biotechnol*, 2004. **22**(1): pp. 47–52.
59. Bruchez, M., Jr., *et al.*, Semiconductor nanocrystals as fluorescent biological labels. *Science*, 1998. **281**(5385): pp. 2013–2016.

60. Michalet, X., *et al.*, Quantum dots for live cells, *in vivo* imaging, and diagnostics. *Science*, 2005. **307**(5709): pp. 538–544.

61. Gao, X., *et al.*, *In vivo* cancer targeting and imaging with semiconductor quantum dots. *Nat Biotechnol*, 2004. **22**(8): pp. 969–976.

62. Medintz, I.L., *et al.*, Quantum dot bioconjugates for imaging, labelling and sensing. *Nat Mater*, 2005. **4**(6): pp. 435–446.

63. Jiang, G., *et al.*, Target specific intracellular delivery of siRNA/PEI-HA complex by receptor mediated endocytosis. *Mol Pharm*, 2009. **6**(3): pp. 727–737.

64. Chen, A.A., *et al.*, Quantum dots to monitor RNAi delivery and improve gene silencing. *Nucleic Acids Res*, 2005. **33**(22): p. e190.

65. Cole, A.J., V.C. Yang, and A.E. David, Cancer theranostics: the rise of targeted magnetic nanoparticles. *Trends Biotechnol*, 2011. **29**(7): pp. 323–332.

66. Medarova, Z., *et al.*, *In vivo* imaging of siRNA delivery and silencing in tumors. *Nat Med*, 2007. **13**(3): pp. 372–377.

67. Kumar, M., *et al.*, Image-guided breast tumor therapy using a small interfering RNA nanodrug. *Cancer Res*, 2010. **70**(19): pp. 7553–7561.

68. Jal, P.K., S. Patel, and B.K. Mishra, Chemical modification of silica surface by immobilization of functional groups for extractive concentration of metal ions. *Talanta*, 2004. **62**(5): pp. 1005–1028.

69. Slowing, II, B.G. Trewyn, and V.S. Lin, Mesoporous silica nanoparticles for intracellular delivery of membrane-impermeable proteins. *J Am Chem Soc*, 2007. **129**(28): pp. 8845–8849.

70. Kneuer, C., *et al.*, A nonviral DNA delivery system based on surface modified silica-nanoparticles can efficiently transfect cells *in vitro*. *Bioconjug Chem*, 2000. **11**(6): pp. 926–932.

71. Rosenholm, J.M., C. Sahlgren, and M. Linden, Towards multifunctional, targeted drug delivery systems using mesoporous silica nanoparticles — opportunities & challenges. *Nanoscale*, 2010. **2**(10): pp. 1870–1883.

72. Hu-Lieskovan, S., *et al.*, Sequence-specific knockdown of EWS-FLI1 by targeted, nonviral delivery of small interfering RNA inhibits tumor growth in a murine model of metastatic Ewing's sarcoma. *Cancer Res*, 2005. **65**(19): pp. 8984–8992.

73. Pirollo, K.F., *et al.*, Materializing the potential of small interfering RNA via a tumor-targeting nanodelivery system. *Cancer Res*, 2007. **67**(7): pp. 2938–2943.

74. Takahashi, Y., *et al.*, Gene silencing in primary and metastatic tumors by small interfering RNA delivery in mice: quantitative analysis using

melanoma cells expressing firefly and sea pansy luciferases. *J Control Release*, 2005. **105**(3): pp. 332–343.

75. Takeshita, F., *et al.*, Efficient delivery of small interfering RNA to bone-metastatic tumors by using atelocollagen *in vivo*. *Proc Natl Acad Sci USA*, 2005. **102**(34): pp. 12177–12182.

76. Sah, D.W., Therapeutic potential of RNA interference for neurological disorders. *Life Sci*, 2006. **79**(19): pp. 1773–1780.

77. Elbashir, S.M., *et al.*, Duplexes of 21-nucleotide RNAs mediate RNA interference in cultured mammalian cells. *Nature*, 2001. **411**(6836): pp. 494–498.

78. de Fougerolles, A., *et al.*, RNA interference *in vivo*: toward synthetic small inhibitory RNA-based therapeutics. *Methods Enzymol*, 2005. **392**: pp. 278–296.

79. Schiffelers, R.M., *et al.*, Cancer siRNA therapy by tumor selective delivery with ligand-targeted sterically stabilized nanoparticle. *Nucleic Acids Res*, 2004. **32**(19): p. e149.

80. Li, S.D., and L. Huang, Surface-modified LPD nanoparticles for tumor targeting. *Ann N Y Acad Sci*, 2006. **1082**: pp. 1–8.

81. Li, S.D., and L. Huang, Nanoparticles evading the reticuloendothelial system: role of the supported bilayer. *Biochim Biophys Acta*, 2009. **1788**(10): pp. 2259–2266.

82. Abra, R.M., M.E. Bosworth, and C.A. Hunt, Liposome disposition *in vivo*: effects of pre-dosing with lipsomes. *Res Commun Chem Pathol Pharmacol*, 1980. **29**(2): pp. 349–360.

83. Moghimi, S.M., and S.S. Davis, Innovations in avoiding particle clearance from blood by Kupffer cells: cause for reflection. *Crit Rev Ther Drug Carrier Syst*, 1994. **11**(1): pp. 31–59.

84. Chrastina, A., K.A. Massey, and J.E. Schnitzer, Overcoming *in vivo* barriers to targeted nanodelivery. *Wiley Interdiscip Rev Nanomed Nanobiotechnol*, 2011. **3**(4): pp. 421–437.

85. Uster, P.S., *et al.*, Insertion of poly(ethylene glycol) derivatized phospholipid into pre-formed liposomes results in prolonged *in vivo* circulation time. *FEBS Lett*, 1996. **386**(2–3): pp. 243–246.

86. Yu, B., *et al.*, Receptor-targeted nanocarriers for therapeutic delivery to cancer. *Mol Membr Biol*, 2010. **27**(7): pp. 286–298.

87. Templeton, N.S., *et al.*, Improved DNA: liposome complexes for increased systemic delivery and gene expression. *Nat Biotechnol*, 1997. **15**(7): pp. 647–652.

88. Hong, K., *et al.*, Stabilization of cationic liposome-plasmid DNA complexes by polyamines and poly(ethylene glycol)-phospholipid conjugates for efficient *in vivo* gene delivery. *FEBS Lett*, 1997. **400**(2): pp. 233–237.

89. Liu, Y., *et al.*, Factors influencing the efficiency of cationic liposome-mediated intravenous gene delivery. *Nat Biotechnol*, 1997. **15**(2): pp. 167–173.

90. Semple, S.C., A. Chonn, and P.R. Cullis, Influence of cholesterol on the association of plasma proteins with liposomes. *Biochemistry*, 1996. **35**(8): pp. 2521–2525.

91. Crook, K., *et al.*, Inclusion of cholesterol in DOTAP transfection complexes increases the delivery of DNA to cells *in vitro* in the presence of serum. *Gene Ther*, 1998. **5**(1): pp. 137–143.

92. Li, S.D., *et al.*, Tumor-targeted delivery of siRNA by self-assembled nanoparticles. *Mol Ther*, 2008. **16**(1): pp. 163–169.

93. Bolcato-Bellemin, A.L., *et al.*, Sticky overhangs enhance siRNA-mediated gene silencing. *Proc Natl Acad Sci USA*, 2007. **104**(41): pp. 16050–16055.

94. Li, S., and L. Huang, *In vivo* gene transfer via intravenous administration of cationic lipid-protamine-DNA (LPD) complexes. *Gene Ther*, 1997. **4**(9): pp. 891–900.

95. Li, S.D., and L. Huang, Targeted delivery of antisense oligodeoxynucleotide and small interference RNA into lung cancer cells. *Mol Pharm*, 2006. **3**(5): pp. 579–588.

96. Silvander, M., M. Johnsson, and K. Edwards, Effects of PEG-lipids on permeability of phosphatidylcholine/cholesterol liposomes in buffer and in human serum. *Chem Phys Lipids*, 1998. **97**(1): pp. 15–26.

97. Brannon-Peppas, L., and J.O. Blanchette, Nanoparticle and targeted systems for cancer therapy. *Adv Drug Deliv Rev*, 2004. **56**(11): pp. 1649–1659.

98. John, C.S., *et al.*, Targeting sigma receptor-binding benzamides as *in vivo* diagnostic and therapeutic agents for human prostate tumors. *Cancer Res*, 1999. **59**(18): pp. 4578–4583.

99. John, C.S., *et al.*, Synthesis, *in vitro* pharmacologic characterization, and preclinical evaluation of N-[2-(1′-piperidinyl)ethyl]-3-[125I]iodo-4-methoxybenzamide (P[125I]MBA) for imaging breast cancer. *Nucl Med Biol*, 1999. **26**(4): pp. 377–382.

100. John, C.S., *et al.*, A malignant melanoma imaging agent: synthesis, characterization, *in vitro* binding and biodistribution of iodine-125-(2-piperidinylaminoethyl)4-iodobenzamide. *J Nucl Med*, 1993. **34**(12): pp. 2169–2175.

101. John, C.S., *et al.*, Synthesis, *in vitro* validation and *in vivo* pharmacokinetics of [125I]N-[2-(4-iodophenyl)ethyl]-N-methyl-2-(1-piperidinyl) ethylamine: a high-affinity ligand for imaging sigma receptor positive tumors. *Nucl Med Biol*, 1996. **23**(6): pp. 761–766.

102. John, C.S., B.J. Vilner, and W.D. Bowen, Synthesis and characterization of [125I]-N-(N-benzylpiperidin-4-yl)-4- iodobenzamide, a new sigma receptor radiopharmaceutical: high-affinity binding to MCF-7 breast tumor cells. *J Med Chem*, 1994. **37**(12): pp. 1737–1739.

103. John, C.S., *et al.*, Synthesis and pharmacological characterization of 4-[125I]-N-(N-benzylpiperidin-4-yl)-4-iodobenzamide: a high affinity sigma receptor ligand for potential imaging of breast cancer. *Cancer Res*, 1995. **55**(14): pp. 3022–3027.

104. Vilner, B.J., C.S. John, and W.D. Bowen, Sigma-1 and sigma-2 receptors are expressed in a wide variety of human and rodent tumor cell lines. *Cancer Res*, 1995. **55**(2): pp. 408–413.

105. Walker, J.M., *et al.*, Sigma receptors: biology and function. *Pharmacol Rev*, 1990. **42**(4): pp. 355–402.

106. Banerjee, R., *et al.*, Anisamide-targeted stealth liposomes: a potent carrier for targeting doxorubicin to human prostate cancer cells. *Int J Cancer*, 2004. **112**(4): pp. 693–700.

107. Li, S.D., S. Chono, and L. Huang, Efficient gene silencing in metastatic tumor by siRNA formulated in surface-modified nanoparticles. *J Control Release*, 2008. **126**(1): pp. 77–84.

108. Zhu, X., *et al.*, Identification of internalizing human single-chain antibodies targeting brain tumor sphere cells. *Mol Cancer Ther*, 2010. **9**(7): pp. 2131–2141.

109. Pirollo, K.F., and E.H. Chang, Does a targeting ligand influence nanoparticle tumor localization or uptake? *Trends Biotechnol*, 2008. **26**(10): pp. 552–558.

110. Bartlett, D.W., *et al.*, Impact of tumor-specific targeting on the biodistribution and efficacy of siRNA nanoparticles measured by multimodality *in vivo* imaging. *Proc Natl Acad Sci USA*, 2007. **104**(39): pp. 15549–15554.

111. Choi, C.H., *et al.*, Mechanism of active targeting in solid tumors with transferrin-containing gold nanoparticles. *Proc Natl Acad Sci USA*, 2010. **107**(3): pp. 1235–1240.

112. Huang, X., *et al.*, A reexamination of active and passive tumor targeting by using rod-shaped gold nanocrystals and covalently conjugated peptide ligands. *ACS Nano*, 2010. **4**(10): pp. 5887–5896.

113. Kirpotin, D.B., *et al.*, Antibody targeting of long-circulating lipidic nanoparticles does not increase tumor localization but does increase internalization in animal models. *Cancer Res*, 2006. **66**(13): pp. 6732–6740.

114. Kappler, M., *et al.*, Increased survivin transcript levels: an independent negative predictor of survival in soft tissue sarcoma patients. *Int J Cancer*, 2001. **95**(6): pp. 360–363.

115. Kato, J., *et al.*, Expression of survivin in esophageal cancer: correlation with the prognosis and response to chemotherapy. *Int J Cancer*, 2001. **95**(2): pp. 92–95.

116. Asanuma, K., *et al.*, Survivin as a radioresistance factor in pancreatic cancer. *Jpn J Cancer Res*, 2000. **91**(11): pp. 1204–1209.

117. Byers, L.A., and J.V. Heymach, Dual targeting of the vascular endothelial growth factor and epidermal growth factor receptor pathways: rationale and clinical applications for non-small-cell lung cancer. *Clin Lung Cancer*, 2007. **8**(Suppl. 2): pp. S79–85.

118. Dassonville, O., *et al.*, EGFR targeting therapies: monoclonal antibodies versus tyrosine kinase inhibitors. Similarities and differences. *Crit Rev Oncol Hematol*, 2007. **62**(1): pp. 53–61.

119. Sathornsumetee, S., and J.N. Rich, New approaches to primary brain tumor treatment. *Anticancer Drugs*, 2006. **17**(9): pp. 1003–1016.

120. Wong, S.F., Cetuximab: an epidermal growth factor receptor monoclonal antibody for the treatment of colorectal cancer. *Clin Ther*, 2005. **27**(6): pp. 684–694.

121. Fan, Q.W., and W.A. Weiss, RNA interference against a glioma-derived allele of EGFR induces blockade at G2M. *Oncogene*, 2005. **24**(5): pp. 829–837.

122. Kang, C.S., *et al.*, An *in vitro* study on the suppressive effect of glioma cell growth induced by plasmid-based small interference RNA (siRNA) targeting human epidermal growth factor receptor. *J Neurooncol*, 2005. **74**(3): pp. 267–273.

123. Kang, C.S., *et al.*, Suppression of EGFR expression by antisense or small interference RNA inhibits U251 glioma cell growth *in vitro* and *in vivo*. *Cancer Gene Ther*, 2006. **13**(5): pp. 530–538.

124. Vollmann, A., *et al.*, Effective silencing of EGFR with RNAi demonstrates non-EGFR dependent proliferation of glioma cells. *Int J Oncol*, 2006. **28**(6): pp. 1531–1542.

125. Wu, X., *et al.*, Combining siRNAs at two different sites in the EGFR to suppress its expression, induce apoptosis, and enhance 5-fluorouracil sensitivity of colon cancer cells. *J Surg Res*, 2007. **138**(1): pp. 56–63.

126. Zhang, X., *et al.*, Tumor growth inhibition by simultaneously blocking epidermal growth factor receptor and cyclooxygenase-2 in a xenograft model. *Clin Cancer Res*, 2005. **11**(17): pp. 6261–6269.

127. Song, E., *et al.*, Antibody mediated *in vivo* delivery of small interfering RNAs via cell-surface receptors. *Nat Biotechnol*, 2005. **23**(6): pp. 709–717.

128. Halaby, M.J., and D.Q. Yang, p53 translational control: a new facet of p53 regulation and its implication for tumorigenesis and cancer therapeutics. *Gene*, 2007. **395**(1–2): pp. 1–7.

129. de Nigris, F., M.L. Balestrieri, and C. Napoli, Targeting c-Myc, Ras and IGF cascade to treat cancer and vascular disorders. *Cell Cycle*, 2006. **5**(15): pp. 1621–1628.

130. Grothey, A., Future directions in vascular endothelial growth factor-targeted therapy for metastatic colorectal cancer. *Semin Oncol*, 2006. **33**(5 Suppl. 10): pp. S41–49.

131. Li, S.D., S. Chono, and L. Huang, Efficient oncogene silencing and metastasis inhibition via systemic delivery of siRNA. *Mol Ther*, 2008. **16**(5): pp. 942–946.

132. Xu, D.H., J.Q. Gao, and W.Q. Liang, Liposome-based intracellular kinetics of doxorubicin in K562/DOX cells. *Pharmazie*, 2008. **63**(9): pp. 646–649.

133. Chen, Y., *et al.*, Novel cationic lipid that delivers siRNA and enhances therapeutic effect in lung cancer cells. *Mol Pharm*, 2009. **6**(3): pp. 696–705.

134. Chen, Y., *et al.*, Multifunctional nanoparticles delivering small interfering RNA and doxorubicin overcome drug resistance in cancer. *J Biol Chem*, 2010. **285**(29): pp. 22639–22650.

135. Nakamura, Y., H. Sato, and T. Motokura, Development of multidrug resistance due to multiple factors including P-glycoprotein

overexpression under K-selection after MYC and HRAS oncogene activation. *Int J Cancer*, 2006. **118**(10): pp. 2448–2454.

136. He, Y., *et al.*, The role of c-myc in regulating mdr1 gene expression in tumor cell line KB. *Chin Med J (Engl)*, 2000. **113**(9): pp. 848–851.

137. Banerjee, S., *et al.*, Down-regulation of ras and myc expression associated with mdr-1 overexpression in adriamycin-resistant tumor cells. *Cell Mol Biol*, 1992. **38**(6): pp. 561–570.

138. Duceppe, N., and M. Tabrizian, Factors influencing the transfection efficiency of ultra low molecular weight chitosan/hyaluronic acid nanoparticles. *Biomaterials*, 2009. **30**(13): pp. 2625–2631.

139. Chono, S., *et al.*, An efficient and low immunostimulatory nanoparticle formulation for systemic siRNA delivery to the tumor. *J Control Release*, 2008. **131**(1): pp. 64–69.

140. Schwartz, D.A., *et al.*, CpG motifs in bacterial DNA cause inflammation in the lower respiratory tract. *J Clin Invest*, 1997. **100**(1): pp. 68–73.

141. Lapcik, L., Jr., *et al.*, Hyaluronan: preparation, structure, properties, and applications. *Chem Rev*, 1998. **98**(8): pp. 2663–2684.

142. Toole, B.P., Hyaluronan: from extracellular glue to pericellular cue. *Nat Rev Cancer*, 2004. **4**(7): pp. 528–539.

143. Eliaz, R.E., and F.C. Szoka, Jr., Liposome-encapsulated doxorubicin targeted to CD44: a strategy to kill CD44-overexpressing tumor cells. *Cancer Res*, 2001. **61**(6): pp. 2592–2601.

144. Platt, V.M., and F.C. Szoka, Jr., Anticancer therapeutics: targeting macromolecules and nanocarriers to hyaluronan or CD44, a hyaluronan receptor. *Mol Pharm*, 2008. **5**(4): pp. 474–486.

145. Uskokovic, V., and D.P. Uskokovic, Nanosized hydroxyapatite and other calcium phosphates: chemistry of formation and application as drug and gene delivery agents. *J Biomed Mater Res B Appl Biomater*, 2011. **96**(1): pp. 152–191.

146. Auzenne, E., *et al.*, Hyaluronic acid-paclitaxel: antitumor efficacy against CD44(+) human ovarian carcinoma xenografts. *Neoplasia*, 2007. **9**(6): pp. 479–486.

147. Li, J., *et al.*, Biodegradable calcium phosphate nanoparticle with lipid coating for systemic siRNA delivery. *J Control Release*, 2010. **142**(3): pp. 416–421.

148. Kester, M., *et al.*, Calcium phosphate nanocomposite particles for *in vitro* imaging and encapsulated chemotherapeutic drug delivery to cancer cells. *Nano Lett*, 2008. **8**(12): pp. 4116–4121.

149. Olton, D., *et al.*, Nanostructured calcium phosphates (NanoCaPs) for non-viral gene delivery: influence of the synthesis parameters on transfection efficiency. *Biomaterials*, 2007. **28**(6): pp. 1267–1279.

150. Singh, N., *et al.*, NanoGenotoxicology: the DNA damaging potential of engineered nanomaterials. *Biomaterials*, 2009. **30**(23–24): pp. 3891–3914.

151. Li, J., Y. Yang, and L. Huang, Calcium phosphate nanoparticles with an asymmetric lipid bilayer coating for siRNA delivery to the tumor. *J Control Release*, 2012. **158**: pp. 108–114 (NIHMS335072).

152. Cotmore, J.M., G. Nichols, Jr., and R.E. Wuthier, Phospholipid-calcium phosphate complex: enhanced calcium migration in the presence of phosphate. *Science*, 1971. **172**(3990): pp. 1339–1341.

153. Xu, Y., and F.C. Szoka, Jr., Mechanism of DNA release from cationic liposome/DNA complexes used in cell transfection. *Biochemistry*, 1996. **35**(18): pp. 5616–5623.

154. Hafez, I.M., N. Maurer, and P.R. Cullis, On the mechanism whereby cationic lipids promote intracellular delivery of polynucleic acids. *Gene Ther*, 2001. **8**(15): pp. 1188–1196.

155. Yang, Y., *et al.*, Systemic delivery of siRNA via LCP nanoparticle efficiently inhibits lung metastasis. *Mol Ther*, 2012. **20**(3): pp. 609–615.

Chapter 4

Delivery of Single siRNA Molecules

Caroline Palm-Apergi and Steven F. Dowdy*

Howard Hughes Medical Institute, Department of Cellular and Molecular Medicine, UCSD School of Medicine, 9500 Gilman Drive, La Jolla, CA 92093-0686, USA

*sdowdy@ucsd.edu

The major impediment in development of macromolecular drug therapeutics, including peptides, proteins, and siRNAs, is delivery of the drug into the cytoplasm or nucleus of cells where it will exert its biological effect. Macromolecular siRNAs are too large, too charged, and too hydrophilic to cross the hydrophobic plasma membrane, which serves as a protective barrier to protect keep foreign molecules and pathogens out of the cell. Thus, cellular delivery of siRNAs remains a significant barrier and rate-limiting step for development of siRNA therapeutics. Current, siRNA delivery approaches rely on condensing and packaging siRNAs into ~100 nm liposome or polymer nanoparticles. However, lipid nanoparticles are 100 megaDaltons in size, roughly 5,000-fold larger than the 14 kDaltons siRNA being delivered, which severely limits the diffusion coefficient, pharmacokinetics, and potential of lipid nanoparticle therapeutics. An alternative siRNA delivery approach is to deliver one siRNA into a cell at a time using Peptide Transduction Domains (PTDs) delivery

Nanotechnology for the Delivery of Therapeutic Nucleic Acids
Edited by Dan Peer

ISBN 978-981-4411-04-2 (Hardcover), 978-981-4411-05-9 (eBook)
www.panstanford.com

peptides, also called Cell-Penetrating Peptides (CPPs). However, due to the negative charge of the siRNA, conjugation of positively charged PTDs results in aggregation, cytotoxicity, and poor delivery into cells. This chapter describes a new delivery strategy that utilizes a PTD fusion protein with a double-stranded RNA-binding domain (DRBD) that binds to and masks the siRNA negative charge, allowing for PTD-mediated delivery into cells and induction of RNAi responses.

4.1 Introduction

The hydrophobic plasma membrane of cells pose a significant barrier in drug delivery of macromolecular therapeutics that need to cross it in order to reach the cytoplasm or nucleus to exert its intended therapeutic effect. Several transporters in the plasma membrane are able to internalize small molecule therapeutics; however, for larger macromolecules such as peptides, proteins, and oligonucleotides, a delivery vehicle is required to transport the cargo across the cellular membrane. Moreover, many important drug targets are difficult to inhibit with small molecule drugs and monoclonal antibodies. One of the most promising macromolecular drug candidates, which are able to target the undruggable genome, is short interfering RNAs (siRNAs). The 21 nucleotide siRNA duplex is hydrophilic, negatively charged and can induce a natural cellular process called RNA interference (RNAi) once it has reached inside the cytoplasm. The process of RNAi was discovered over a decade ago, and because of its high catalytic activity and target selectivity, RNAi has given new hope for modern medicine to battle a wide range of diseases (Fire *et al.*, 1998). However, suitable delivery vehicles have been difficult to find because of its polyanionic nature. Although formulations such as nanoparticles and polymers have been used both *in vitro* and *in vivo*, unfortunately, many of them have suffered from toxicity and poor efficiency (Pecot *et al.*, 2011). The problems involved in delivering siRNA into cells have slowed down the progress of developing siRNA into a drug. Therefore, the number one priority for RNAi has been to find a delivery vehicle that can transport the active siRNA molecule across the cell membrane in a non-toxic manner.

4.1.1 RNA Interference

RNAi is a post-transcriptional gene-silencing mechanism that utilizes siRNA to bind and induce degradation of complementary mRNA, consequently, leading to a decrease of the corresponding protein. It is a natural cellular process initiated by Dicer, a ribonuclease III enzyme that cleaves endogenous long double-stranded RNA into 21–23 mer RNA duplexes with 5′-phosphate groups and 2-nucleotide 3′ overhangs. After Dicer has cleaved, the duplex is unwound and one of the strands, the antisense strand also called the guide strand, is loaded into the RNA-induced silencing complex (RISC). One of the RISC proteins, the human immunodeficiency virus (HIV)-1 transactivating response RNA-binding protein (TRBP), then recruits another protein to the complex, Argonaute 2 (Ago2) that subsequently cleaves the complementary passenger strand (Frank *et al.*, 2010, Chendrimada *et al.*, 2005). A crucial part of the guide strand responsible for recognizing the mRNA is called the seed sequence and corresponds to nucleotides 2–8 starting from the 5′ end. The RISC-guide strand complex employs the seed sequence as a template to find and bind complementary mRNAs and directs Ago2 to cleave the mRNA between nucleotides 10–11 from the 5′ end of the guide strand.

Once the mRNA is cleaved, the whole process is repeated as the RISC-guide strand complex detects and cleaves new mRNA strands, continuously. Thus, RNAi is an extremely efficient process since one guide strand alone can cleave a high number of complementary mRNA, resulting in a substantial decrease of the corresponding protein. Importantly, it has been shown that the RNAi process can be hijacked by introducing exogenous, synthetic 21 nucleotide siRNAs, to inhibit the production of specific target proteins in cultured mammalian cells (Elbashir *et al.*, 2001). Moreover, compared to small molecule drugs, it is a preferable treatment strategy to use against many diseases, since it is possible to target any kind of pathological protein. Due to the sequence-specificity of this process, a new field of therapeutic drug design has opened up.

4.1.2 Modification of siRNAs

Many tissues, in contrast to local targets, can only be reached by systemic administration. Therefore, a wide range of modifications has been applied to the siRNA molecule, in order to overcome the obstacles that siRNA formulations face when they are administered via the bloodstream. Before the siRNA reaches the cytoplasm of its target cell, it needs to avoid many challenges such as phagocytic uptake, kidney filtration, and enzymatic degradation by nucleases

(Alexis *et al.*, 2008). In addition, naked siRNA can trigger the immune system through toll-like receptors (TLRs). Thus, several modifications have been applied to the siRNA molecule in order to improve its stability against nuclease degradation and to evade activation of the immune response. For instance, introduction of phosphorothioates on the backbone of the siRNA, as well as 2′-O-methylpurines and 2′-fluoropyrimidines on the ribose molecule, results in resistance against nuclease degradation (Czauderna *et al.*, 2003). Furthermore, it has been shown that immune stimulation by synthetic siRNA can be completely abrogated by selective incorporation of 2′-O-methyl uridine and guanosine on the ribose, into one strand of the siRNA duplex (Morrissey *et al.*, 2005, Judge *et al.*, 2006).

4.1.3 Off-Target Effects

Another major impediment in RNAi therapeutics has been off-target effects. Both the guide and passenger strands can induce off-target side effects that lead to measurable phenotypes in unwanted targets. Initially, it was suggested that as few as 11 contiguous nucleotides were enough to detect cross-reactions with non-targeted genes of limited sequence similarity (Jackson *et al.*, 2003). However, more recent studies have shown that a perfect match between the 3′ untranslated region (UTR) of the mRNA and the seed sequence of the siRNA guide strand is required to induce off-target effects (Birmingham *et al.*, 2006). However, if the seed sequence of the guide strand matches the 5′ UTR or the open reading frame (ORF) of the mRNA, the off-target effects do not seem to occur. Moreover, by modifying a single nucleotide in the seed sequence to a 2′-O-methyl, off-targeting activity can be reduced significantly (Robbins *et al.*, 2007). Up to date, no occurrence of seed-sequence-mediated off-target effects has been reported in clinical trials of siRNA (Kim and Rossi, 2007; Davidson and McCray, 2011). Although big progress has been made with siRNA stability and toxicity, delivery is still the biggest problem to be solved.

4.2 Delivery of siRNA

4.2.1 Peptide Transduction Domains

More than 20 years ago, during the development of an activity assay, it was discovered that the HIV Trans-Activator of Transcription

(TAT) protein was able to enter the cytoplasm of cells without any transporter (Frankel and Pabo, 1988). At the same time, in another laboratory, chemically synthesized TAT was also shown to enter cells rapidly (Green and Loewenstein, 1988). Some years later, it was found that a 60-amino-acids-long peptide from another transcription factor, corresponding to the homeobox of the Antennadapedia protein in *Drosophila*, entered cells efficiently (Joliot *et al.*, 1991). This initiated a new field of delivery vehicles named PTDs or CPPs that has evolved and expanded greatly over the last decades. The PTDs are short cationic peptides approximately 10–30 amino acids long and can deliver different cargoes into cells, such as drugs and oligonucleotides, both *in vitro* and *in vivo* (Snyder and Dowdy, 2004, Eguchi *et al.*, 2009). Moreover, several PTDs are now being analyzed in clinical trials, as reviewed elsewhere (van den Berg and Dowdy, 2011).

4.2.2 Delivery of siRNA-PTD Nanoparticles

The uptake mechanism of PTDs has been studied intensively and several pathways of endocytosis have been suggested as well as direct penetration through the membrane (Kaplan *et al.*, 2005; Palm-Apergi *et al.*; 2009, Ferrari *et al.*; 2003, Richard *et al.*, 2005). It has also been shown that different forms of endocytosis can function simultaneously, and by blocking one pathway, other pathways may become more active. Furthermore, the uptake is dependent on both concentration and cargo. Several studies have utilized PTDs to deliver siRNA into the cytoplasm, both as covalently linked conjugates and electrostatically bound nanoparticles. However, the efficiency of covalently conjugated PTD-siRNA has been low due to the precipitation and aggregation that occur when the cationic residues of the PTD interact with the anionic phosphate backbone of the siRNAs (Moschos *et al.*, 2007). In contrast, some successful studies have taken advantage of the negative charges of the siRNA and the positively charged PTD to form electrostatically bound PTD-siRNA nanoparticles to induce knockdown of target mRNA both *in vitro* and *in vivo*.

One of the first studies to utilize the electrostatic interaction between the siRNA and the PTD synthesized a peptide named MPG that consisted of two domains. The first domain was hydrophilic and cationic in order to bind the anionic siRNA, whereas the second domain contained hydrophobic amino acids that engaged the lipids in the plasma membrane to induce internalization and subsequently RNAi (Morris *et al.*, 1997). This non-covalent interaction strategy continued in development over the years, and in the most recent report, a new PTD, CADY, was introduced, which consists of aromatic tryptophan and cationic arginine residues. Upon binding to the plasma membrane, CADY adopts a helical conformation, with one side exposing the charged arginine residues to bind the siRNAs and the other side exposing the hydrophobic residues to favor cellular uptake. Incubation of siRNAs with CADY PTDs resulted in nanoparticles that were able to induce RNAi in mammalian cells (Crombez *et al.*, 2009). Another study incorporated a trifluoromethylquinoline moiety covalently attached to a stearylated-TP10 delivery peptide, named PepFect6, in order to increase the release of nanoparticles trapped in endosomes (Andaloussi *et al.*, 2011). The trifluoromethylquinoline moiety functioned similarly to chloroquine in that it buffered the endosomes to increase the release of siRNA-PepFect6 nanoparticles from the vesicles. The PepFect6-siRNA nanoparticles showed efficient RNAi responses both in tissue culture cells and in kidney, lung, and liver, after systemic delivery with no apparent toxicity.

Although there are successful PTD nanoparticle delivery strategies, they may still pose a problem from a drug point of view due to their enormous size that usually differs from particle to particle and due to potential toxicity issues; it would be preferable to deliver the siRNA as a single molecule with a defined size, rather than a siRNA-nanoparticle with varying size. In order to deliver the siRNA as a single molecule, the negative charges on the siRNA must be masked to prevent neutralization of the PTD and inhibition of its delivery properties before PTD conjugation can occur.

4.2.3 RNA Binding Proteins

RNA-binding proteins (RBPs) play important roles in post-transcriptional control and several RNA-regulated processes, such as splicing, transport, stability, localization, and translation. Most RBPs have a RNA-binding domain that recognizes specific mRNA sequence motifs or secondary structures, whereas some recognize common structural features. In general, RBPs bind regulatory sequences that are usually located in 5′ UTR and/or 3′ UTR of the mRNAs to exert post-transcriptional regulation (Lee and Schedl, 2006). Approximately 500 RBPs have been identified, and one of them is the diverse class of double-stranded RNA binding proteins (DRBP) that contain certain domains able to bind double-stranded RNA. These double-stranded RNA binding domains (DRBDs) are highly abundant and can be found in many different species. The DRBPs have different functions but are similar in that they contain varying copies of DRBDs.

A consensus sequence of about 65–68 amino acid residues specific for binding dsRNA was identified in 1992 (St Johnston *et al.,* 1992). It was later shown that a minimum number of 16 base pairs of siRNAs are required for binding, however, for longer RNAs, only 11 base pairs are required (Bevilacqua and Cech, 1996). Moreover, DRBDs are known to bind the compressed A-form of the siRNA sequence independently and almost without changing the structure. The DRBD binds the 2′-hydroxyl groups on the sugar phosphate backbone, and every 11 mer is involved in the interaction. Consequently, DRBDs cannot bind dsDNA or even RNA/DNA hybrids, since dsDNA is predominantly found in the more open B-form (Bevilacqua and Cech, 1996). Three regions are responsible for binding the dsRNA, where two out of three bind the minor groove, whereas the third binds the major groove. One of the most commonly used PTDs belongs to the HIV-1 and is derived from the TAT protein (Vivès *et al.,* 1997). By designing a fusion protein of three TAT peptides and a DRBD (PTD-DRBD), the previous obstacles in siRNA delivery can be overcome (Eguchi *et al.,* 2009, Palm-Apergi *et al.,* 2011).

4.2.4 Delivery of Single siRNA Molecules by PTD-DRBD

Initially, while developing this new delivery strategy, several PTDs were assessed for their ability to deliver siRNA. The goal was to conjugate a single PTD to a single siRNA molecule. However, the positive charges of the PTD and the negative charges on the siRNA lead to precipitation, aggregation, and charge neutralization during each conjugation attempt, resulting in an inactivated PTD. Those observations lead to the conclusion that the anionic phosphodiester backbone of the siRNA needs to be masked, before any PTD conjugation can occur without any precipitation. Since several TAT fusion proteins had been developed to study cellular mechanisms (Nagahara *et al.,* 1998), it seemed natural to design a fusion protein with a PTD delivery domain and a DRBD, able to bind and neutralize the anionic charges on the siRNA. Since the DRBD is small and contains around 65 amino acid residues that specifically bind about 12–16 base pairs of the double-stranded RNA backbone, it can bind on 90° surface quadrants of the double-stranded RNA helix, resulting in four DRBDs encompassing a single siRNA (Ryter and Schultz, 1998). Each PTD-DRBD fusion protein contains three TAT sequences and one single DRBD that binds the siRNA with a $K_D \sim 10^{-9}$, thereby masking the negative charges of the siRNA. Thus, the positively charged side chains of the PTD delivery vehicle can interact with the plasma membrane and induce internalization.

The DRBD binds the 2′-hydroxyl groups of the siRNA specifically and enables the PTD-DRBD fusion protein to deliver the siRNA into the cytoplasm of cell, resulting in efficient knockdown of target mRNA in a non-cytotoxic manner. Several primary cell lines that are known to be difficult to transfect have shown efficient target protein inhibition after PTD-DRBD delivery of siRNA (Eguchi *et al.,* 2009). Since siRNAs have been shown to stimulate for instance TLRs, it was important to investigate possible innate immune response activation. Although the siRNA needs to have 2′-hydroxyl groups in order for the DRBD to bind, the PTD-DRBD-mediated delivery of siRNA does not activate Interferon-α nor TNF-α responses in primary peripheral blood mononuclear cells. Moreover, intranasal PTD-DRBD delivery of siRNA in transgenic ROSA26 mice, stably expressing tissue restricted luciferase, showed extensive reduction of luciferase expression, demonstrating that the PTD-DRBD-mediated delivery of siRNA is specific and induce RNAi responses in reporter mouse models *in vivo*. The induction of *in vivo* synthetic lethal RNAi responses, mediated by PTD-DRBD delivered siRNA, was also confirmed in glioblastoma mouse models (Michiue *et al.,* 2009). Thus, the PTD-DRBD-delivered siRNA strategy is non-toxic and applicable to both *in vitro* and *in vivo* models. Importantly, the DRBD binds the siRNA independently of sequence composition, prevents activation of the innate immune response, and serves as an excellent vehicle for siRNA delivery together with the PTD.

4.3 Discussion

Although there are several benefits with nanoparticle-based delivery of siRNA, there are, in addition to size, some disadvantages such as off-target effects and cytotoxicity that have been reported and need to be addressed (Pecot *et al.,* 2011). In nanoparticle delivery of siRNA, the negative charge of the siRNA is utilized to its advantage, as the nanoparticle is produced by complexing the cationic residues of the PTD to the anionic siRNA, electrostatically. Furthermore, some unwanted characteristics of siRNA, such as RNAse-mediated degradation and TLR-mediated innate immune responses, are overcome by encapsulating the siRNA molecule in nanoparticle formulations. The negatively charged phosphodiester backbone of the siRNA easily binds cationic lipids, peptides, and polymers electrostatically, resulting in nanoparticles of varying sizes of around 1–1000 nm. After the siRNA has been administered systemically, it needs to bind and cross the vascular endothelial barrier during the bloodstream circulation. Most molecules larger than 5 nm will remain in circulation until they are cleared from the body and do not generally cross the capillary endothelium (Whitehead *et al.*, 2009). However, some tissues, such as liver, spleen, and certain tumors, allow passage of larger molecules, whereas molecules less than 50 kDa in size are usually cleared through the kidneys. Indeed, phosphorothioate oligonucleotides have been shown to have a plasma clearance half-life of only 5 minutes (Rappaport *et al.*, 1995).

Intravenously injected particles with a diameter bigger than 100 nm have a high probability of becoming trapped in the reticuloendothelial system in the liver, spleen, lung, and bone marrow, followed by macrophage and monocyte degradation (Pecot *et al.*, 2011). At the same time, a large-sized nanoparticle will, due to the size restriction of kidney filtration, have an increased circulation time following systemic administration that will increase the chance of engaging the desired target to further benefit their pharmacokinetic properties. However, the physical and chemical properties of the surface can lead to hemolysis, thrombogenicity, and complement activation that can cause alterations in biodistribution and potential toxicity (Dobrovolskaia *et al.*, 2008). Furthermore, both cationic lipids and polymers known to be toxic, due to the cationic charge ratio that is required for packaging and the unnatural composition of the polymer or lipid, may very well engage the lipids in the plasma membrane of the cell, resulting in cellular toxicity.

In addition to the plasma membrane, there are other extracellular factors in the microenvironment that may need to be overcome. For instance, the siRNA could be prematurely released from the nanoparticle because of the chemical and physical properties of the extracellular matrix (Burke and Pun, 2008). Moreover, the diffusion constant of nanoparticles through the interstitial spaces of tissues is dramatically reduced because of their relatively large size compared to PTD-DRBD-delivered siRNA. This is supported by the numerous studies where systemically administered nanoparticle-mediated siRNA is targeted to the liver (Meade and Dowdy, 2009, Whitehead *et al.*, 2009). Therefore, it has been important to focus on developing an alternative strategy to deliver siRNA as a single molecule that would benefit both its potency and pharmacokinetic properties, when delivered into tissue culture cells or mouse models *in vivo*.

4.4 Conclusions

In this chapter, we have described an alternative delivery approach that circumvents general nanoparticle strategies. Although many challenges in siRNA delivery can be overcome by nanoparticle encapsulation, the most obvious disadvantage compared to the PTD-DRBD delivery vehicle is the size. Lipid nanoparticles are about 100 mDa in size compared to the small PTD-DRBD that is only 16 kDa. Furthermore, DRBDs bind siRNA specifically, sequence independently, and mask the negative charges of the siRNA, leaving the positively charged PTD domain free to bind the plasma membrane and induce internalization. Once inside the cytoplasm, the siRNA is able to induce a rapid and long-lasting RNAi response in a non-cytotoxic fashion both *in vitro* and *in vivo*. In summary, the ability to induce synthetic RNAi with PTD-DRBD-delivered siRNA in mouse models holds great promise for future RNAi therapeutic development.

Acknowledgments

C. P.-A. was funded by a Knut & Alice Wallenberg's Foundation and the Sweden-America Foundation Fellowship. This work was supported by the Leukemia and Lymphoma Society, the Pardee Foundation, California Institute of Regenerative Medicine, and Howard Hughes Medical Institute.

References

1. Alexis, F., Pridgen, E., Molnar, L. K., and Farokhzad, O. C. (2008). Factors affecting the clearance and biodistribution of polymeric nanoparticles, *Mol. Pharm.* **5**, pp. 505–515.

2. Andaloussi, S. E., Lehto, T., Mäger, I., Rosenthal-Aizman, K., Oprea, I. I., Simonson, O. E., *et al.* (2011). Design of a peptide-based vector, PepFect6, for efficient delivery of siRNA in cell culture and systemically *in vivo*, *Nucleic Acids Res.* **39**, pp. 3972–3987.
3. Bevilacqua, P. C., and Cech, T. R. (1996). Minor-groove recognition of double-stranded RNA by the double-stranded RNA-binding domain from the RNA-activated protein kinase PKR, *Biochemistry.* **35**, pp. 9983–9984.
4. Birmingham, A., Anderson, E. M., Reynolds, A., Ilsley-Tyree, D., Leake, D., Fedorov. Y., *et al.* (2006). 3′ UTR seed matches, but not overall identity, are associated with RNAi off-targets, *Nat. Methods.* **3**, pp. 199–204.
5. Burke, R. S., and Pun, S. H. (2008). Extracellular barriers to *in vivo* PEI and PEGylated PEI polyplex-mediated gene delivery to the liver, *Bioconjug. Chem.* **19**, pp. 693–704.
6. Chendrimada, T. P., Gregory, R. I., Kumaraswamy, E., Norman, J., Cooch, N., Nishikura, K., *et al.* (2005). TRBP recruits the Dicer complex to Ago2 for microRNA processing and gene silencing, *Nature Lett.* **436**, pp. 740–744.
7. Crombez, L., Aldrian-Herrada, G., Konate, K., Nguyen, Q. N., McMaster, G. K., Brasseur, R., Heitz, F., Divita, G. (2009). A new potent secondary amphipathic cell-penetrating peptide for siRNA delivery into mammalian cells, *Mol. Ther.* **17**, pp. 95–103.
8. Czauderna, F., Fechtner, M., Dames, S., Aygün, H., Klippel, A., Pronk, G. J., *et al.* (2003). Structural variations and stabilising modifications of synthetic siRNAs in mammalian cells, *Nucleic Acids Res.* **31**, pp. 2705–2716.
9. Davidson, B. L., and McCray, P. B. Jr. (2011). Current prospects for RNA interference-based therapies, *Nat. Rev. Genet.* **12**, pp. 329–340.
10. Dobrovolskaia, M. A., Aggarwal, P., Hall, J. B., and McNeil, S. E. (2008). Preclinical studies to understand nanoparticle interaction with the immune system and its potential effects on nanoparticle biodistribution, *Mol. Pharm.* **5**, pp. 487–495.
11. Eguchi, A., Meade, B. R., Chang, Y. C., Fredrickson, C. T., Willert, K., Puri, N., *et al.* (2009). Efficient siRNA delivery into primary cells by a peptide transduction domain-dsRNA binding domain fusion protein, *Nat. Biotechnol.* **27**, pp. 567–571.
12. Elbashir, S. M., Harborth, J., Lendeckel, W., Yalcin, A., Weber, K., and Tuschl, T. (2001). Duplexes of 21-nucleotide RNAs mediate RNA interference in cultured mammalian cells, *Nature.* **411**, pp. 494–498.

13. Ferrari, A., Pellegrini, V., Arcangeli, C., Fittipaldi, A., Giacca, M., and Beltram, F. (2003). Caveolae-mediated internalization of extracellular HIV-1 tat fusion proteins visualized in real time, *J. Biol. Chem.* **278**, pp. 34141–34149.

14. Fire, A., Xu, S., Montgomery, M. K., Kostas, S. A., Driver, S. E., and Mello, C. C. (1998). Potent and specific genetic interference by double stranded RNA in Caenorhabditis elegans, *Nature.* **391**, pp. 806–811.

15. Frank, F., Sonenberg, N., and Nagar, B. (2010). Structural basis for 5′-nucleotide base-specific recognition of guide RNA by human Ago2, *Nature.* **465**, pp. 818–822.

16. Frankel, A. D., and Pabo, C. O. (1988). Cellular uptake of the tat protein from human immunodeficiency virus, *Cell.* **55**, pp. 1189–1193.

17. Green, M., and Loewenstein, P. M. (1988). Autonomous functional domains of chemically synthesized human immunodeficiency virus tat trans-activator protein, *Cell.* **55**, pp. 1179–1188.

18. Jackson, A. L., Bartz, S. R., Schelter, J., Kobayashi, S. V., Burchard, J., Mao, M., *et al.* (2003). Expression profiling reveals off-target gene regulation by RNAi, *Nat. Biotechnol.* **21**, pp. 635–637.

19. Joliot, A., Pernelle, C., Deagostini-Bazin, H., and Prochiantz, A. (1991). Antennapedia homeobox peptide regulates neural morphogenesis, *Proc. Natl. Acad. Sci.* **88**, pp. 1864–1868.

20. Judge, A. D., Bola, G., Lee, A. C., and MacLachlan, I. (2006). Design of noninflammatory synthetic siRNA mediating potent gene silencing *in vivo*, *Mol. Ther.* **13**, pp. 494–505.

21. Kaplan, I. M., Wadia, J. S., and Dowdy, S. F. (2005). Cationic TAT peptide transduction domain enters cells by macropinocytosis, *J. Control Release.* **102**, pp. 247–253.

22. Kim, D. H., and Rossi, J. J. (2007). Strategies for silencing human disease using RNA interference, *Nat. Rev. Genet.* **8**, pp. 173–184.

23. Lee, M. H., and Schedl, T. (2006). RNA-binding proteins, *WormBook.* **18**, pp. 1–13.

24. Meade, B. R., and Dowdy, S. F. (2009). The road to therapeutic RNA interference (RNAi): tackling the 800 pound siRNA delivery gorilla, *Discov. Med.* **8**, pp. 253–256.

25. Michiue, H., Eguchi, A., Scadeng, M., and Dowdy, S. F. (2009). Induction of *in vivo* synthetic lethal RNAi responses to treat glioblastoma, *Cancer Biol. Ther.* **8**, pp. 2306–2313.

26. Morris, M. C., Vidal, P., Chaloin, L., Heitz, F., and Divita, G. (1997). A new peptide vector for efficient delivery of oligonucleotides into mammalian cells, *Nucleic Acids Res.* **25**, pp. 2730–2736.

27. Morrissey, D. V., Lockridge, J. A., Shaw, L., Blanchard, K., Jensen, K., Breen, W., *et al.* (2005). Potent and persistent *in vivo* anti-HBV activity of chemically modified siRNAs, *Nat. Biotechnol.* **23**, pp. 1002–1007.

28. Moschos, S. A., Jones, S. W., Perry, M. M., Williams, A. E., Erjefalt, J. S., Turner, J. J., *et al.* (2007). Lung delivery studies using siRNA conjugated to TAT(48–60) and penetratin reveal peptide induced reduction in gene expression and induction of innate immunity, *Bioconjug. Chem.* **18**, pp. 1450–1459.

29. Nagahara, H., Vocero-Akbani, A. M., Snyderm E. L., Ho, A., Latham, D. G., Lissy, N. A., *et al.* (1998). Transduction of full-length TAT fusion proteins into mammalian cells: TAT-p27Kip1 induces cell migration, *Nat. Med.* **4**, pp. 1449–1452.

30. Palm-Apergi, C., Lorents, A., Padari, K., Pooga, M., and Hällbrink, M. (2009). The membrane repair response masks membrane disturbances caused by cell-penetrating peptide uptake, *FASEB J.* **23**, pp. 214–223.

31. Palm-Apergi, C., Eguchi, A., and Dowdy, S. F. (2011). PTD-DRBD siRNA delivery, *Methods Mol. Biol.* **683**, pp. 339–347.

32. Pecot, C. V., Calin, G. A., Coleman, R. L., Lopez-Berestein, G., and Sood, A. K. (2011). RNA interference in the clinic: challenges and future directions, *Nat. Rev. Cancer.* **11**, pp. 59–67.

33. Rappaport, J., Hanss, B., Kopp, J. B., Copeland, T. D., Bruggeman, L. A., Coffman, T. M., *et al.* (1995). Transport of phosphorothioate oligonucleotides in kidney: implications for molecular therapy, *Kidney Int.* **47**, pp. 1462–1469.

34. Richard, J. P., Melikov, K., Brooks, H., Prevot, P., Lebleu, B., Chernomordik, L. V., *et al.* (2005). Cellular uptake of unconjugated TAT peptide involves clathrin-dependent endocytosis and heparan sulfate receptors, *J. Biol. Chem.* **280**, pp. 15300–15306.

35. Robbins, M., Judge, A., Liang, L., McClintock, K., Yaworski, E., and MacLachlan, I. (2007). 2′-O-Methyl-modified RNAs act as TLR7 antagonists, *Mol. Ther.* **15**, pp. 1663–1669.

36. Ryter, J. M., and Schultz, S. C. (1998). Molecular basis of double-stranded RNA-protein interactions: structure of a dsRNA-binding domain complexed with dsRNA, *EMBO J.* **17**, pp. 7505–7513.

37. Snyder, E. L., and Dowdy, S. F. (2004). Cell penetrating peptides in drug delivery, *Pharm. Res.* **21**, pp. 389–393.

38. St Johnston, D., Brown, N. H., Gall, J. G., and Jantsch, M. (1992). A conserved double-stranded RNA-binding domain, *Proc. Natl. Acad. Sci.* **89**, pp. 10979–10983.

39. van den Berg, A., and Dowdy, S. F. (2011). Protein transduction domain delivery of therapeutic macromolecules, *Curr. Opin. Biotechnol.* **22**, pp. 888–893.

40. Vivès, E., Brodin, P., and Lebleu, B. (1997). A truncated HIV-1 Tat protein basic domain rapidly translocates through the plasma membrane and accumulates in the cell nucleus, *J. Biol. Chem.* **272**, pp. 16010–16017.

41. Whitehead, K. A., Langer, R., and Anderson, D. G. (2009). Knocking down barriers: advances in siRNA delivery, *Nat. Rev. Drug Discov.* **8**, pp. 129–138.

Chapter 5

Cell-Specific Aptamer-Functionalized RNAi: A New Prospect for Targeted siRNA Delivery

Jiehua Zhou and John J. Rossi*

Molecular and Cellular Biology, Beckman Research Institute of City of Hope, 1500 E Duarte Road, Duarte, CA 91010, USA

*jrossi@coh.org

In recent years, nucleic acid based therapeutics has progressed rapidly. In addition to their utility as a stand-alone strategy, these therapeutic nucleic acids that include small interfering (si)RNAs, microRNAs, aptamers, antisense DNAs, DNAzymes, and ribozymes can potentially be used as adjuvants for multipronged drug treatment. In particular, the highly specific mechanism of RNA interference (RNAi) which inhibits the expression of disease genes is extremely attractive for the treatment of a wide variety of human maladies. However, one of the most formidable impediments to clinical translation of RNAi is effective delivery of the small interfering RNAs (siRNAs) to specific cells or tissues at therapeutic doses. Currently, extensive efforts to develop various internalizing molecules have been made for target-specific siRNA delivery. Among them, cell-specific aptamers represent an exciting approach for targeted siRNA delivery. In this chapter, we review the current advances of cell-

Nanotechnology for the Delivery of Therapeutic Nucleic Acids
Edited by Dan Peer

ISBN 978-981-4411-04-2 (Hardcover), 978-981-4411-05-9 (eBook)
www.panstanford.com

specific aptamers in cell recognition and targeted delivery, with a particular focus on the development of the aptamer-functionalized siRNA or nanocarrier for targeted gene silencing.

5.1 Introduction

Although the concept of nucleic acid based therapeutics is not a new one, it has captured the wide attention of scientists for a long time [1, 2]. Over the past few decades, a wide variety of nucleic acids such as siRNAs [1, 2], microRNAs [3, 4], aptamers [5, 6], antisense DNA and RNA [7, 8], mRNAs [9, 10], plasmid DNAs [11], DNAzymes, and ribozymes [12] have been proposed as possible drugs. Some of these nucleic acid candidates have been put on the back burner or are still in development because of current technical challenges; however, some nucleic acids have entered clinic trails, and a few have even successfully made it to market; e.g., Vitravene [13], the antisense oligonucleotide inhibiting cytomegalovirus (CMV)-induced retinitis, and Pegaptanib (or Macugen) [14], the first therapeutic aptamer approved for the treatment of neovascular age-related macular degeneration (AMD).

Since the discovery of RNAi in 1998 [15], the use of RNAi-based therapeutics (for example, siRNAs [16]) to silence target genes associated with human diseases has progressed tremendously due to their high specificity and potency. The research on RNAi has fuelled excitement for their potential clinical application [1]. So far, there are over 20 RNAi-based drugs in early- or mid-stage clinical trials, several of which are indicating strong promise for further drug development [3]. For example, in 2004 Acuity Pharmaceuticals (now Opko Health) announced the first siRNA-related clinical trial, in which Bevasiranib, an unmodified siRNA targeting vascular endothelial growth factor (VEGF), was given in patients with wet AMD [17]. Additionally, the first in-human Phase I clinical trial using a targeted nanoparticle system to deliver siRNA to patients was conducted and showed the direct evidence for siRNA-mediated gene silencing *via* the RNAi pathway [18]. However, like any other drugs in development, some impediments have caused failures in the clinical translation of RNAi-based therapeutics. In March 2009, Opko decided to terminate the Phase III clinical trial of bevasiranib, since it was unlikely to achieve its primary endpoint of reducing vision loss. In this case, as an unmodified siRNA, bevasiranib was given by intravitreal injection without a particular formulation that could enhance its silencing performance. Since negatively charged nucleic acids do not readily traverse cellular membranes and are vulnerable to degradation without some protective covering or/and appropriate chemical modifications, such direct administration ultimately resulted in poor pharmacokinetics and indirect gene silencing. In this regard, some optimized strategies should be considered, which could include appropriate chemical backbone modifications, an effective delivery vehicle, a new dosing schedule and a combinatorial formulation with other therapeutic nucleic acids, and so on.

While this frustration tempered the early excitement in RNAi technology, it triggered extensive efforts to surmount these key hurdles to the widespread use of RNAi as a therapy. Among these hurdles, one of the most important is delivery of RNAi agent (such as siRNAs) to specific cells or tissues at therapeutic doses [19]. With the intent of developing a targeted intracellular delivery system, numerous strategies have been reported, in which various internalizing molecules with high specificity and affinity to a cellular receptor were assembled or complexed with the siRNA. Based on specific interaction between the targeting ligand and its cellular receptor, the recognition and internalization of the therapeutic siRNAs by the target tissue should be capable of dramatic improvement.

Most recently, another type of therapeutic nucleic acids termed aptamers show promise as potential candidates for targeted siRNA delivery. These cell-specific aptamers can actively target a distinct cell population or tissue in a cell-type-specific manner [20, 21]. By functionalizing the cell-specific aptamers to therapeutic agents or delivery vehicles, the cellular uptake is enhanced and the local concentration of the drug in the targeted cells or tissues is increased, thereby improving the therapeutic efficacy. Currently, a number of aptamer-functionalized therapeutics has been successfully developed for targeted delivery [22, 23]. In particular, with the technological maturation and increasing knowledge of RNAi as well as aptamers and their mechanism of action, it seems natural to partner the two therapeutic nucleic acids for expanding the options for targeted RNAi (Fig. 5.1). In this chapter, we review recent progress of cell-specific aptamers that mediate targeted therapy of nucleic acid drugs placing particular attention to development of aptamer-functionalized siRNAs or nanocarrier for targeted gene silencing.

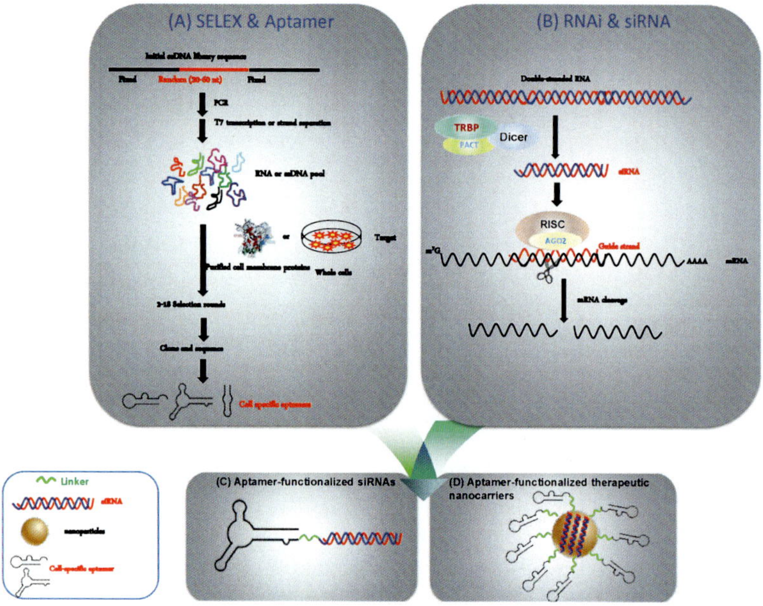

Figure 5.1 Through the combination of SELEX (aptamers) and RNAi (siRNA) techniques, cell-specific aptamer-functionalized RNAi can be achieved. (A) SELEX procedure for aptamer isolation. In order to generate a cell-specific aptamer, the selection procedure can vary from traditional purified protein-based SELEX to whole cell-based SELEX. Generally, in a typical SELEX procedure, the initial single-stranded DNA pool contains a 20-60 nt random sequence, which guarantees the presence of active structures with high binding affinity to the target protein. By repeating selection rounds, aptamers against any given targets can be routinely isolated from an initial combinatorial oligonucleotide library. (B) Mechanisms of RNA interference in mammalian cells. Cytoplasmic double-stranded RNAs (dsRNAs) are processed by a complex consisting of Dicer, TAR RNA-binding protein (TRBP) and protein activator of protein kinase PKR (PACT) into siRNAs, which are loaded into Argonaute 2 (AGO2) and the RNA-induced silencing complex (RISC). The siRNA guide strand recognizes target sites to direct mRNA cleavage, which is carried out by the catalytic domain of AGO2. siRNAs complementary to promoter regions direct transcriptional gene silencing in the nucleus through chromatin changes involving histone methylation (m7G, 7-methylguanylate). (C) Cell-specific aptamer-functionalized siRNA. Cell-internalizing aptamers are covalently fused or non-covalently assembled to siRNA agents. A linker (for example: 2-4-nucleotides, sticky sequence, or biotin-streptavidin connector) connects the aptamer and siRNA. (D) Cell-specific aptamer-functionalized therapeutic nanocarrier. siRNAs are encapsulated into a nanocarrier functionalized by cell-internalizing aptamers.

5.2 Generation of Cell-Specific Aptamers

Developed independently by three research groups in 1990 [24–26], aptamers (from Latin word *aptus* "to fit" and the Greek word *merus* for "piece") are *in vitro* evolved single-stranded nucleic acids that fold into a specific three-dimensional shape to fit together with their targets (Fig. 5.1). During the past 20 years, there has been a rapidly increasing number of publications on the use of aptamers, and the aptamer technique has captured the attention of scientists from various fields, owing to their unique properties [21]. The successful adaption of cell-specific aptamers for targeted therapy has created excitement for generating new, more potent aptamers as well as developing novel and facile selection approaches.

To date, by the use of either traditional purified membrane protein-based SELEX (Systematic Evolution of Ligands by Exponential Enrichment) or whole cell-based SELEX procedures, researchers have successfully isolated an increasing number of cell-specific DNA or RNA aptamers against cell surface biomarker or receptors (Fig. 5.1A). For instance, cell-specific aptamers, raised against PSMA (prostate-specific membrane antigen) [27], CD4 (cluster of differentiation 4) [28], CD30 (also known as TNFRSF8) [29], CD44 [30], HIV-1 glycoprotein gp120 [31, 32], TN-C (tenascin-C) [33], EGFR (epidermal growth factor receptor) [34], PTK7 (protein tyrosine kinase 7) [35], TfR (transferrin receptor) [36], NCL (nucleolin) [37, 38], MUC1 (mucin protein) [39], and PDGF-B (platelet-derived growth factor B) [40], have been successfully harnessed for targeting delivery of various cargoes *in vitro* as well as *in vivo*.

5.2.1 Recombinant Protein-Based SELEX Procedure

In vitro SELEX consists of four main steps: (1) binding to the target; (2) isolation of target-specific species; (3) recovery of target-bound sequences; and (4) re-amplification of recovered species [25]. Through iterative selection rounds, aptamers against any given target can be isolated from an initial combinatorial oligonucleotide library. Numerous successful examples have been reported. In addition to the traditional bead, resin, membrane, and chip-based segmentation approaches, novel, powerful isolation strategies have been developed to accelerate selection procedure and improve efficiency of aptamers, such as, MoloLEX (so called one-step selection) [41], capillary gel electrophoresis [42–44], microfluidic devices [45, 46], and so on. With the automation of SELEX [47, 48], aptamers for multiple targets can be quickly isolated in a high-throughput selection format.

The advantages of the use of the purified protein as target molecule, such as low nonspecific binding and easy control of the selection conditions, allow optimal enrichment of the sequences specific to the target protein [49]. In the majority of cases, cell-specific aptamers have been evolved to target soluble, purified proteins. For example, for the evolution of the anti-PSMA RNA aptamer [27], a purified fusion protein containing a modified extracellular form of PSMA was immobilized on magnetic beads. Therefore, the bound species were readily segmented from unbound sequences *via* a simple washing step, and the bound species were recovered and re-amplified for the next selection round. Similarly, 2′-F-modified RNA aptamers targeting CD4 were generated by immobilizing soluble, recombinant CD4 antigen onto sepharose-beads [28]. Additionally, several 2′-F-modified anti-HIV-1 gp120 RNA aptamers have been successfully identified by using either the BIAcore biosensor system [50, 51] or conventional nitrocellulose filter methods [31]. As the most established isolation method, the cellulose filter-based SELEX has succeeded in generating anti-EGFR aptamers [34] and anti-TfR aptamers [36] highly specific for binding their purified target proteins.

However, in some cases, aptamers raised against the purified target protein were unable to recognize the same target in its native conformation when the protein target is functionally part of a multi-protein complex, is embedded in a physiological context, or is subjected to a post-translational modifications [52]. For example, RNA aptamers that were generated by using the purified extracellular domain of human RTK Ret failed to bind to full-length Ret expressed on the cell surface [53, 54]. Similarly, although RNA aptamers indicated high binding affinity to their target histidine-tagged EGFRvIII ectodomain, they didn't bind to the full-length EGFRvIII protein that is extensively glycosylated when expressed on the cell surface [55].

5.2.2 Whole Cell-Based SELEX Procedure

Recent advances in cell-based SELEX techniques have provided a promising alternative approach for identifying aptamers that can selectively bind to a cell-surface antigen or a particular target cell population [52, 56, 57]. In contrast to the purified protein-based SELEX, the cell-based SELEX method is performed with whole living cells, which ensures the native folded conformation or original glycosylation pattern of the extracellular domains of target proteins. Typically, this procedure includes two essential steps: (1) counter-selection with nontargeted cells that do not express the target molecules on the cell surface; and (2) positive-selection with the targeted cells. By taking advantage of the molecular differences between any two closely related cell populations, it is possible to obtain aptamers against most antigens, even unknown targets or multiprotein complexes if they are specially and selectively expressed on or in the target cells [58].

Recently, several research groups have successfully adopted this strategy to generate cell-specific aptamers. For example, CHO (Chinese hamster ovary) cells that overexpress the recombinant TGF-β (transforming growth factor) type III receptors have been used as a target for the isolation of RNA aptamers against the TGF-β type III receptor [59]. In this case, counter-selection has been conducted with CHO cells lacking the target expression. Another example is a cell-based DNA aptamer able to discriminate rat glioblastoma from microglia cells [60]. Additionally, even with a lack of information about the molecular targets, a panel of cell-specific DNA aptamers has been successfully selected to bind to CCRF-CEM cells [35, 61]. They were capable of distinguishing T-cells and B-cells in patient samples. One of the selected DNA aptamers, sgc8, identified to target PTK7, has been demonstrated to be specifically taken up into lymphoblastic leukemia T-cells *via* a receptor-mediated endocyosis [62].

Although successful in individual cases, one of the most serious drawbacks of cell-based SELEX is the nonspecific binding/uptake of nucleic acids to dead cells, thereby delaying target-specific sequence enrichment, even resulting in the failure of aptamer selection [63, 64]. Unfortunately, during the selection procedure, some steps or conditions (for example, detachment of adherent cells, washing, buffer concentrations, incubation times, etc.) might cause damage or death to some fragile cells. Therefore, the careful elimination of dead cell populations before incubation with aptamer library is crucial and necessary. The complexity of target cell surface markers or the lower expression of the desired targets in cell-based SELEX generally requires a greater number of selection rounds (14 ~ 15 or up to 25) in order to evolve the desired specific sequences. Consequently, some nonspecific sequences might dominate the selection due to the bias of enzymatic amplification reactions.

To overcome these disappointments with the whole cell-based SELEX technology, some optimizations have been made to increase its general applicability. For example, a fluorescence-activated cell sorting (FACS) technology was combined with cell-based SELEX, thus allowing the separation of living cells from dead cells within a cultured cell mixture [64]. Through this novel FACS-SELEX, aptamers specific to vital Burkitt lymphoma B cells were enriched after 10 selection rounds. Alternatively, a gentler method was performed to significantly reduce the proportion of dead cells by dead cell removal microbeads [63]. In addition, high-affinity cell-specific RNA aptamers have been effectively identified with a crossover SELEX strategy [65], in which the targets were crossing from the Tenasin-C(TN-C)-expressing U251 glioblastoma cells (in the initial nine rounds) to the purified TN-C protein (in the last two rounds).

5.3 Cell-Specific Aptamer-Functionalized RNAi

With their favorable properties (such as high affinity and specificity, small physical size, facile *in vitro* evolution, low toxicity, and accessibility for backbone modifications), cell-specific aptamers offer great promise for cell type- or tissue-specific delivery of various therapeutics. For example, RNAi therapeutics (siRNAs), therapeutic nanopaticles, molecular carriers, chemotherapeutics, and molecular imaging probes have been precisely functionalized with cell-specific aptamers. As an actively recognized moiety, aptamers direct specific cell recognition events and consequently facilitate cellular uptake of payloads *via* receptor-mediated endocytosis. Despite displaying "escort" functionality, some cell-specific aptamers possess the ability of directly blocking the function of target proteins or interrupting the disease process, thereby making them attractive platforms to develop multifunctional nanomedicines. The following sections highlight current efforts using cell-specific aptamer to enable targeted delivery of siRNAs (Fig. 5.1C) and their use as therapeutic nanocarriers (Fig. 5.1D).

5.3.1 Cell-Specific Aptamer-Functionalized siRNAs

RNA interference is extensively being harnessed to silence mRNAs encoding pathogenic proteins (Fig. 5.1B,C). The direct application of RNAi-based therapeutic agents, such as 21-22 nt siRNAs or corresponding Dicer substrates (25-27 nt duplexes or small hairpin RNAs), is currently the most common method to harness the RNAi pathway for targeted gene silencing [3]. It has been reported that the silencing potency of an optimal Dicer substrate siRNAs (dsiRNAs) can be greater than for traditional 21-22 nt siRNAs [66]. By using appropriate formulations, targeted delivery of these synthetic RNAi triggers to specific cell populations or tissues can be accomplished with high efficiency. Several representative examples are discussed below.

5.3.1.1 PSMA RNA aptamer-functionalized siRNAs

The first example of a cell-specific apatamer used for targeted siRNA delivery is the RNA aptamer that binds PSMA, a trans-membrane protein highly expressed in human prostate cancer and the vascular endothelium [67, 68]. Through either covalent conjugation or physical assembly, different siRNA molecules have been successfully functionalized with anti-PSMA aptamers to achieve targeted RNAi efficacy. In 2006, Chu *et al.* [69] showed a proof of concept study, in which two-biotinylated anti-PSMA aptamers (A-9) and two-biotinylated 27-mer dsiRNAs targeting lamin A/C or GAPDH were noncovalently assembled on a streptavidin connector *via* biotin-streptavidin interaction. The resulting multivalent construct displayed selective cellular uptake into the cultured PSMA-positive cells and mediated specific RNAi activity.

In the same year, a different approach [70], in which a 2′-F-modified anti-PSMA aptamer (A-10) was covalently appended to the sense strand of a 21-mer siRNA portion, which subsequently was annealed with its complementary antisense strand. This simple covalent aptamer-siRNA chimeric RNA allowed effective PSMA-mediated internalization along with siRNA-triggered gene silencing in athymic mice following intratumoral injection. In a recent effort [71], the same group has further optimized the previous design through rational modifications of the aptamer and siRNA portions. Such optimized second-generation chimeras had increased circulation and bio-availability *in vivo*, resulting in pronounced regression of PSMA-expressing tumors following systemic aptamer-siRNA conjugate administration. By using a similar strategy, Wullner *et al.* [72] generated bivalent anti-PSMA aptamer-siRNA chimeras, further promoting internalization of chimeras and their therapeutic potential. Most recently, Gliboa and his coworkers [73] also demonstrated that the NMD (nonsense-mediated messenger RNA decay) factor targeting siRNAs could be selectively targeted to tumor cells using anti-PSMA aptamer-siRNA chimeras, resulting in significant inhibition of tumor growth after systemic administration in both subcutaneous and metastatic tumor models.

In addition to siRNA or dsiRNA, a short hairpin (sh)RNA specific to DNAPK (DNA-activated protein kinase) was covalently fused to the anti-PSMA aptamer (A-10) and selectively reduced DNAPK expression in PCa (PSMA-positive) cell xenografts and human prostate tissues [74]. It was shown that radiation therapy for locally advanced PCa was specifically improved when the aptamer-shRNA chimeras were combined with ionizing radiation (IR).

5.3.1.2 HIV gp120 RNA aptamer-functionalized siRNAs

The HIV-1 envelop glycoprotein gp120 is exposed on the surface of virus particles and the plasma membrane of HIV-1 infected cells [75]. The interaction of HIV-1 gp120 with the cellular CD4 receptor is crucial step in the entry of HIV into T-cells [76]. HIV-1 gp120 therefore represents a potential therapeutic target to block HIV-1 infection [77]. Recently, our group successfully used 2′-F-modified anti-HIV-1 gp120 RNA aptamers for cell type-specific delivery of anti-HIV-1 dsiRNAs in cultured cells [31, 78] and in HIV-1 infected humanized Rag-hu mice [79]. In our system, the anti-gp120 aptamers had dual functions acting as targeted siRNA delivery agents and as HIV-1 inhibitors. We developed two different types of aptamer-siRNA conjugates: One is a covalent aptamer-siRNA chimera; another one is a noncovalent aptamer-stick-siRNA conjugate in which the aptamer and siRNA portions are joined *via* hybridization using a GC-rich sticky bridge sequence. The aptamer-siRNA conjugates were selectively internalized into HIV-1 infected cells, resulting in several logs of inhibition of HIV-1 replication in the humanized mouse model following systemic administration. Our results also demonstrated that the aptamer anti-HIV-1 *tat/rev* dsiRNA was processed by Dicer, resulting in site specific cleavage of the target mRNA *in vivo*.

Since rapid emergence of viral escape mutants often abrogates the RNAi efficacy, continued efforts have been made to develop an aptamer-mediated combinatorial multi-targeting RNAi approach. In this regard, the anti-gp120 aptamer was conjugated with three different dsiRNAs targeting HIV-1 *tat/rev* transcripts and HIV-1 host dependency factors (CD4 and Transportin-3 (TNPO3)) through the "sticky" bridge [31], thus providing marked suppression of viral loads and protection of $CD4^+$ T-cells *in vivo* (Zhou *et al.*, Functional *in vivo* delivery of multiplexed anti-HIV-1 siRNAs via a chemically synthesized aptamer with a sticky bridge, *Gene Therapy*, 2012, submitted). The combination of aptamer and siRNAs has the added potential benefit of multiple therapeutic modes of action thereby minimizing the potential for viral escape by mutation.

5.3.1.3 CD4 RNA aptamer-functionalized siRNAs

CD4, the primary receptor for the human immunodeficiency virus (HIV), plays an important role in HIV entry into host T-cells [76,80,81]. The anti-CD4 RNA aptamers have provided HIV inhibition in a CD4-specific manner [28]. Most recently, a CD4 aptamer was fused with a siRNA targeting CCR5 and shown to specifically suppress gene expression in $CD4^+$ T-cells and macrophages in cervicovaginal tissue explants [82]. Moreover, the aptamer-siRNA chimeras (CD4-AsiCs) did not activate lymphocytes or stimulate innate immunity. When applied intravaginally to humanized mice, the CD4-AsiCs protected against HIV vaginal transmission. CD4-AsiCs could be used as the active ingredient of a microbicide to prevent HIV sexual transmission.

5.3.2 Cell-Specific Aptamer-Functionalized Therapeutic Nanocarriers

The advent of nanotechnology has greatly accelerated the development of drug delivery, providing a large variety of nanocarriers for disease therapy [83] (Fig. 5.1D). In particular, a multifunctional therapeutic nanocarrier system in which several desirable functions such as therapeutics (RNAi therapeutics, chemotherapy agents, or radioactivity agents), targeting (aptamers or other ligands), and imaging (fluorescent dye) are combined in one nanoscale carrier is highly desirable. As reported recently, several cell-specific aptamers have been successfully assembled with therapeutic nanocarriers for targeted RNAi.

5.3.2.1 CD4 RNA aptamer-functionalized pRNA-nanoparticles

The anti-CD4 RNA aptamers also have been fused into a multifunctional nanoparticle for targeted siRNA delivery. Through the interaction of two interlocking loops, bacterial virus phi29 DNA packaging RNA (pRNA) molecules are able to form dimers, trimers, hexamers, and patterned superstructures [84]. Guo and his co-workers have creatively taken advantage of this self-assembly property to fabricate pRNA-based nanoparticles with a variety of structures and shapes [85]. As an ideal RNA building block for bottoms-up assembly, pRNAs can be covalently fused with therapeutic agents, targeting ligands, or imaging agents and subsequently self-assemble into multifunctional nanopaticles, allowing aptamer-directed cell-specific gene silencing and drug tracking. In contrast to the average size of a normal single siRNA molecule that is well below 10 nm, aptamer-functionalized pRNA nanoparticles (dimers or trimers) have sizes ranging between 20 and 40 nm, which have improved circulation time and biodistribution *in vivo*.

5.3.2.2 PSMA RNA aptamer-functionalized polymer nanocarriers

During the chemical synthesis of nucleic acids, various functional groups can be readily incorporated into the aptamers to functionalize them as nanocarriers. For example, synthetic aptamers containing a 5′-NH_2 group can be conjugated with a terminal carboxylic acid on the surface of nanoparticles using carbodiimide coupling chemistry. Using this method Farokhzad and colleagues [86] reported the first example of a PSMA aptamer-functionalized poly(lactic acid)-block-polyethylene glycol copolymer (PLA-b-PEG nanoparticles) for targeted delivery of rhodamine-labeled dextran. After that, various nanoparticles, including PLGA-b-PEG [87–89] and super-paramagnetic iron oxide [90], have been functionalized with cell-specific aptamers.

Recently, a 5′-NH_2-modified PSMA aptamer was chemically coupled with a branched polyethyleneimine-grafted polyethylene glycol polymer (PEI-PEG) that serves as a vehicle for shRNA delivery [91]. Moreover, aptamers generally have a single 5′-CG-3′ sequence in its predicted double-stranded stem region, which is a preferred binding site for the anthracycline class of anticancer drugs such as doxorubicin (Dox). Therefore, Dox was subsequently intercalated into the aptamer portion. The resulting aptamer-conjugated polyplexes efficiently targeted a shRNA against the anti-apoptotic gene (Bcl-xl) and the anti-cancer drug (Dox) to prostate cancers, which synergistically induced selective cell death of prostate cancer. This combinatorial formulation that was loaded with aptamer, shRNA, and Dox significantly increased the cytotoxicity compared to treatment where the therapeutics is simply mixed.

5.3.2.3 CD30 RNA aptamer-functionalized polymer nanocarriers

CD30, a cell membrane protein of the tumor necrosis factor receptor family, has long been recognized as an important, unique biomarker on many lymphomas of diverse origin and as activation molecule on B- and T-cells [92]. ALCL (anaplastic large cell lymphoma) cells, an aggressive T-cell lymphoma, exhibit an abnormal expression of the ALK (anaplastic lymphoma kinase) oncogene and unique surface expression of CD30 [93]. Similar to a CD30-specific antibody, the CD30 RNA aptamer is able to selectively bind to intact CD30-expressing lymphoma cells [29, 94]. Most recently, a nanocomplex was formulated by incorporating both an ALK siRNA and a CD30 RNA aptamer onto nano-sized polyethyleneimine-citrate carriers *via* noncovalent interaction [95]. Exposure of ALCL cells to this targeted nanocomplexes, with a peak hydrodynamic diameter of ~140 nm, specifically silenced ALK gene expression and induced growth arrest and apoptosis.

5.4 Conclusions and Perspectives

With the promise of exquisite specificity and high efficacy, the prospect of using nucleic-acid molecules for treating human malignancies and other diseases, especially using RNAi-based therapeutics for targeted gene silencing, remains a tantalizing goal. Although conceptually elegant, some setbacks in early clinical trials have dampened the luster of RNAi therapeutics. The main issues are as follows: (1) inefficient delivery to the correct cells or tissues; (2) suboptimal gene silencing activity, owing to instability and poor bioavailability of RNAi agents; (3) toxicities due to off-target effects or immune activation. Among these issues, efficient delivery remains the primary challenge. Therefore, delivery formulations must be developed to overcome this key hurdle before RNAi-based therapies can be realized as a general approach for the treatment of human diseases.

The power of aptamers as siRNA and drug delivery vehicles is prompting the widespread development of multifunctional formulation for targeted RNAi therapeutics. Advantages of a cell type specific, aptamer-functionalized nanocarrier system are as follows: (1) a nanoscale size that not only allows preferential accumulation in the tumor tissue/organ in the passive targeting mode but also improves circulation time and biodistribution; (2) cell type-specific aptamers facilitate selective internalization in target cells; (3) a combinatorial therapy format that simultaneously administers two or more therapeutic agents to achieve synergistic therapeutic efficacy; (4) fluorescent labeling allows real-time tracking of drugs for monitoring therapeutic responses. Despite some of the successful examples reviewed in this chapter, continued efforts are still required to further refine these formulations for their clinical translation. It is still important to explore the molecular events associated with RNAi and aptamer-mediated target recognition, thereby accelerating the development of aptamer-siRNAs as diagnostic and therapeutic agents.

References

1. Davidson, BL, and McCray, PB, Jr. (2011). Current prospects for RNA interference-based therapies. *Nat Rev Genet*, **12**. pp. 329–340.
2. Whitehead, KA, Langer, R, and Anderson, DG (2009). Knocking down barriers: advances in siRNA delivery. *Nat Rev Drug Discov*, **8**. pp. 129–138.
3. Lares, MR, Rossi, JJ, and Ouellet, DL (2010). RNAi and small interfering RNAs in human disease therapeutic applications. *Trends Biotechnol*, **28**. pp. 570–579.
4. Small, EM, and Olson, EN (2011). Pervasive roles of microRNAs in cardiovascular biology. *Nature*, **469**. pp. 336–342.
5. Keefe, AD, Pai, S, and Ellington, A (2010). Aptamers as therapeutics. *Nat Rev Drug Discov*, **9**. pp. 537–550.
6. Bunka, DH, Platonova, O, and Stockley, PG (2010). Development of aptamer therapeutics. *Curr Opin Pharmacol*, **10**. pp. 557–562.

7. Stephenson, ML, and Zamecnik, PC (1978). Inhibition of Rous sarcoma viral RNA translation by a specific oligodeoxyribonucleotide. *Proc Natl Acad Sci USA*, **75**. pp. 285–288.

8. Kurreck, J (2003). Antisense technologies. Improvement through novel chemical modifications. *Eur J Biochem*, **270**. pp. 1628–1644.

9. Warren, L, *et al.* (2010). Highly efficient reprogramming to pluripotency and directed differentiation of human cells with synthetic modified mRNA. *Cell Stem Cell*, **7**. pp. 618–630.

10. Kormann, MS, *et al.* (2011). Expression of therapeutic proteins after delivery of chemically modified mRNA in mice. *Nat Biotechnol*, **29**. pp. 154–157.

11. Luo, D, and Saltzman, WM (2000). Synthetic DNA delivery systems. *Nat Biotechnol*, **18**. pp. 33–37.

12. DiGiusto, DL, *et al.* (2010). RNA-based gene therapy for HIV with lentiviral vector-modified CD34(+) cells in patients undergoing transplantation for AIDS-related lymphoma. *Sci Transl Med*, **2**. pp. 36–43.

13. Prasad, SC, and Roy, I (2008). Nucleic acid based therapeutic molecules. *Curr Res Inf Pharm Sci*, **9**. p. 7.

14. Ng, EW, Shima, DT, Calias, P, Cunningham, ET, Jr., Guyer, DR, and Adamis, AP (2006). Pegaptanib, a targeted anti-VEGF aptamer for ocular vascular disease. *Nat Rev Drug Discov*, **5**. pp. 123–132.

15. Fire, A, Xu, S, Montgomery, MK, Kostas, SA, Driver, SE, and Mello, CC (1998). Potent and specific genetic interference by double-stranded RNA in Caenorhabditis elegans. *Nature*, **391**. pp. 806–811.

16. Zamore, PD, Tuschl, T, Sharp, PA, and Bartel, DP (2000). RNAi: double-stranded RNA directs the ATP-dependent cleavage of mRNA at 21 to 23 nucleotide intervals. *Cell*, **101**. pp. 25–33.

17. Singerman, L (2009). Combination therapy using the small interfering RNA bevasiranib. *Retina*, **29**. pp. S49–50.

18. Davis, ME, *et al.* (2010). Evidence of RNAi in humans from systemically administered siRNA via targeted nanoparticles. *Nature*, **464**. pp. 1067–1070.

19. Castanotto, D, and Rossi, JJ (2009). The promises and pitfalls of RNA-interference-based therapeutics. *Nature*, **457**. pp. 426–433.

20. Mayer, G (2009). The chemical biology of aptamers. *Angew Chem Int Ed Engl*, **48**. pp. 2672–2689.

21. Syed, MA, and Pervaiz, S (2010). Advances in aptamers. *Oligonucleotides*, **20**. pp. 215–224.

22. Zhou, J, and Rossi, JJ (2009). The therapeutic potential of cell-internalizing aptamers. *Curr Top Med Chem*, **9**. pp. 1144–1157.
23. Zhou, J, and Rossi, JJ (2011). Cell-specific aptamer-mediated targeted drug delivery. *Oligonucleotides*, **21**. pp. 1–10.
24. Ellington, AD, and Szostak, JW (1990). *In vitro* selection of RNA molecules that bind specific ligands. *Nature*, **346**. pp. 818–822.
25. Tuerk, C, and Gold, L (1990). Systematic evolution of ligands by exponential enrichment: RNA ligands to bacteriophage T4 DNA polymerase. *Science*, **249**. pp. 505–510.
26. Robertson, DL, and Joyce, GF (1990). Selection *in vitro* of an RNA enzyme that specifically cleaves single-stranded DNA. *Nature*, **344**. pp. 467–468.
27. Lupold, SE, Hicke, BJ, Lin, Y, and Coffey, DS (2002). Identification and characterization of nuclease-stabilized RNA molecules that bind human prostate cancer cells via the prostate-specific membrane antigen. *Cancer Res*, **62**. pp. 4029–4033.
28. Kraus, E, James, W, and Barclay, AN (1998). Cutting edge: novel RNA ligands able to bind CD4 antigen and inhibit CD4+ T lymphocyte function. *J Immunol*, **160**. pp. 5209–5212.
29. Mori, T, Oguro, A, Ohtsu, T, and Nakamura, Y (2004). RNA aptamers selected against the receptor activator of NF-kappaB acquire general affinity to proteins of the tumor necrosis factor receptor family. *Nucleic Acids Res*, **32**. pp. 6120–6128.
30. Somasunderam, A, *et al.* (2010). Combinatorial selection of DNA thioaptamers targeted to the HA binding domain of human CD44. *Biochemistry*, **49**. pp. 9106–9112.
31. Zhou, J, *et al.* (2009). Selection, characterization and application of new RNA HIV gp 120 aptamers for facile delivery of Dicer substrate siRNAs into HIV infected cells. *Nucleic Acids Res*, **37**. pp. 3094–3109
32. Dey, AK, Khati, M, Tang, M, Wyatt, R, Lea, SM, and James, W (2005). An aptamer that neutralizes R5 strains of human immunodeficiency virus type 1 blocks gp120-CCR5 interaction. *J Virol*, **79**. pp. 13806–13810.
33. Hicke, BJ, *et al.* (2006). Tumor targeting by an aptamer. *J Nucl Med*, **47**. pp. 668–678.
34. Li, N, Larson, T, Nguyen, HH, Sokolov, KV, and Ellington, AD (2010). Directed evolution of gold nanoparticle delivery to cells. *Chem Commun (Camb)*, **46**. pp. 392–394.
35. Shangguan, D, *et al.* (2006). Aptamers evolved from live cells as effective molecular probes for cancer study. *Proc Natl Acad Sci USA*, **103**. pp. 11838–11843.

36. Chen, CH, *et al.* (2008). Aptamer-based endocytosis of a lysosomal enzyme. *Proc Natl Acad Sci USA*, **105**. pp. 15908–15913.

37. Soundararajan, S, Chen, W, Spicer, EK, Courtenay-Luck, N, and Fernandes, DJ (2008). The nucleolin targeting aptamer AS1411 destabilizes Bcl-2 messenger RNA in human breast cancer cells. *Cancer Res*, **68**. pp. 2358–2365.

38. Ireson, CR, and Kelland, LR (2006). Discovery and development of anticancer aptamers. *Mol Cancer Ther*, **5**. pp. 2957–2962.

39. Ferreira, CS, Cheung, MC, Missailidis, S, Bisland, S, and Gariepy, J (2009). Phototoxic aptamers selectively enter and kill epithelial cancer cells. *Nucleic Acids Res*, **37**. pp. 866–876.

40. Soontornworajit, B, Zhou, J, Zhang, Z, and Wang, Y (2010). Aptamer-functionalized *in situ* injectable hydrogel for controlled protein release. *Biomacromolecules*, **11**. pp. 2724–2730.

41. Nitsche, A, *et al.* (2007). One-step selection of Vaccinia virus-binding DNA aptamers by MonoLEX. *BMC Biotechnolm*, **7**. p. 48.

42. Berezovski, M, Musheev, M, Drabovich, A, and Krylov, SN (2006). Non-SELEX selection of aptamers. *J Am Chem Soc*, **128**. pp. 1410–1411.

43. Berezovski, M, *et al.* (2005). Nonequilibrium capillary electrophoresis of equilibrium mixtures: a universal tool for development of aptamers. *J Am Chem Soc*, **127**. pp. 3165–3171.

44. Mallikaratchy, P, Stahelin, RV, Cao, Z, Cho, W, and Tan, W (2006). Selection of DNA ligands for protein kinase C-delta. *Chem Commun (Camb)*, pp. 3229–3231.

45. Farokhzad, OC, *et al.* (2005). Microfluidic system for studying the interaction of nanoparticles and microparticles with cells. *Anal Chem*, **77**. pp. 5453–5459.

46. Xu, Y, Phillips, JA, Yan, J, Li, Q, Fan, ZH, and Tan, W (2009). Aptamer-based microfluidic device for enrichment, sorting, and detection of multiple cancer cells. *Anal Chem*, **81**. pp. 7436–7442.

47. Eulberg, D, Buchner, K, Maasch, C, and Klussmann, S (2005). Development of an automated *in vitro* selection protocol to obtain RNA-based aptamers: identification of a biostable substance P antagonist. *Nucleic Acids Res*, **33**. p. e45.

48. Cox, JC, Rudolph, P, and Ellington, AD (1998). Automated RNA selection. *Biotechnol Prog*, **14**. pp. 845–850.

49. Kulbachinskiy, AV (2007). Methods for selection of aptamers to protein targets. *Biochemistry (Mosc)*, **72**. pp. 1505–1518.

50. Khati, M, Schuman, M, Ibrahim, J, Sattentau, Q, Gordon, S, and James, W (2003). Neutralization of infectivity of diverse R5 clinical isolates of human immunodeficiency virus type 1 by gp120-binding 2′F-RNA aptamers. *J Virol*, **77**. pp. 12692–12698.

51. Sayer, N, Ibrahim, J, Turner, K, Tahiri-Alaoui, A, and James, W (2002). Structural characterization of a 2′F-RNA aptamer that binds a HIV-1 SU glycoprotein, gp120. *Biochem Biophys Res Commun*, **293**. pp. 924–931.

52. Cerchia, L, and de Franciscis, V (2010). Targeting cancer cells with nucleic acid aptamers. *Trends Biotechnol*, **28**. pp. 517–525.

53. Cerchia, L, *et al.* (2005). Neutralizing aptamers from whole-cell SELEX inhibit the RET receptor tyrosine kinase. *PLoS Biol*, **3**. p. e123.

54. Pestourie, C, *et al.* (2006). Comparison of different strategies to select aptamers against a transmembrane protein target. *Oligonucleotides*, **16**. pp. 323–335.

55. Liu, Y, *et al.* (2009). Aptamers selected against the unglycosylated EGFRvIII ectodomain and delivered intracellularly reduce membrane-bound EGFRvIII and induce apoptosis. *Biol Chem*, **390**. pp. 137–144.

56. Phillips, JA, Lopez-Colon, D, Zhu, Z, Xu, Y, and Tan, W (2008). Applications of aptamers in cancer cell biology. *Anal Chim Acta*, **621**. pp. 101–108.

57. Guo, KT, Paul, A, Schichor, C, Ziemer, G, and Wendel, HP (2008). Cell-SELEX: novel perspectives of aptamer-based therapeutics. *Int J Mol Sci*, **9**. pp. 668–678.

58. Fang, X, and Tan, W (2009). Aptamers generated from cell-SELEX for molecular medicine: a chemical biology approach. *Acc Chem Res*, **43**. pp. 48–57.

59. Ohuchi, SP, Ohtsu, T, and Nakamura, Y (2006). Selection of RNA aptamers against recombinant transforming growth factor-beta type III receptor displayed on cell surface. *Biochimie*, **88**. pp. 897–904.

60. Blank, M, Weinschenk, T, Priemer, M, and Schluesener, H (2001). Systematic evolution of a DNA aptamer binding to rat brain tumor microvessels. Selective targeting of endothelial regulatory protein pigpen. *J Biol Chem*, **276**. pp. 16464–16468.

61. Shangguan, D, Cao, ZC, Li, Y, and Tan, W (2007). Aptamers evolved from cultured cancer cells reveal molecular differences of cancer cells in patient samples. *Clin Chem*, **53**. pp. 1153–1155.

62. Xiao, Z, Shangguan, D, Cao, Z, Fang, X, and Tan, W (2008). Cell-specific internalization study of an aptamer from whole cell selection. *Chemistry*, **14**. pp. 1769–1775.

63. Avci-Adali, M, Metzger, M, Perle, N, Ziemer, G, and Wendel, HP (2010). Pitfalls of cell-systematic evolution of ligands by exponential enrichment (SELEX): existing dead cells during *in vitro* selection anticipate the enrichment of specific aptamers. *Oligonucleotides*, **20**. pp. 317–323.

64. Raddatz, MS, Dolf, A, Endl, E, Knolle, P, Famulok, M, and Mayer, G (2008). Enrichment of cell-targeting and population-specific aptamers by fluorescence-activated cell sorting. *Angew Chem Int Ed Engl*, **47**. pp. 5190–5193.

65. Hicke, BJ, *et al.* (2001). Tenascin-C aptamers are generated using tumor cells and purified protein. *J Biol Chem*, **276**. pp. 48644–48654.

66. Kim, DH, Behlke, MA, Rose, SD, Chang, MS, Choi, S, and Rossi, JJ (2005). Synthetic dsRNA Dicer substrates enhance RNAi potency and efficacy. *Nat Biotechnol*, **23**. pp. 222–226.

67. Liu, H, *et al.* (1998). Constitutive and antibody-induced internalization of prostate-specific membrane antigen. *Cancer Res*, **58**. pp. 4055–4060.

68. Tasch, J, Gong, M, Sadelain, M, and Heston, WD (2001). A unique folate hydrolase, prostate-specific membrane antigen (PSMA): a target for immunotherapy? *Crit Rev Immunol*, **21**. pp. 249–261.

69. Chu, TC, Twu, KY, Ellington, AD, and Levy, M (2006). Aptamer mediated siRNA delivery. *Nucleic Acids Res*, **34**. p. e73.

70. McNamara, JO, 2nd, *et al.* (2006). Cell type-specific delivery of siRNAs with aptamer-siRNA chimeras. *Nat Biotechnol*, **24**. pp. 1005–1015.

71. Dassie, JP, *et al.* (2009). Systemic administration of optimized aptamer-siRNA chimeras promotes regression of PSMA-expressing tumors. *Nat Biotechnol*, **27**. pp. 839–849.

72. Wullner, U, Neef, I, Eller, A, Kleines, M, Tur, MK, and Barth, S (2008). Cell-specific induction of apoptosis by rationally designed bivalent aptamer-siRNA transcripts silencing eukaryotic elongation factor 2. *Curr Cancer Drug Targets*, **8**. pp. 554–565.

73. Pastor, F, Kolonias, D, Giangrande, PH, and Gilboa, E (2010). Induction of tumour immunity by targeted inhibition of nonsense-mediated mRNA decay. *Nature*, **465**. pp. 227–230.

74. Ni, X, *et al.* (2011). Prostate-targeted radiosensitization via aptamer-shRNA chimeras in human tumor xenografts. *J Clin Invest*, **121**. pp. 2383–2390.

75. Kwong, PD, Wyatt, R, Robinson, J, Sweet, RW, Sodroski, J, and Hendrickson, WA (1998). Structure of an HIV gp120 envelope glycoprotein in complex with the CD4 receptor and a neutralizing human antibody. *Nature*, **393**. pp. 648–659.

76. Hussey, RE, *et al.* (1988). A soluble CD4 protein selectively inhibits HIV replication and syncytium formation. *Nature*, **331**. pp. 78–81.
77. Clayton, R, *et al.* (2007). Binding kinetics, uptake and intracellular accumulation of F105, an anti-gp120 human IgG1kappa monoclonal antibody, in HIV-1 infected cells. *J Virol Methods*, **139**. pp. 17–23.
78. Zhou, J, Li, H, Li, S, Zaia, J, and Rossi, JJ (2008). Novel dual inhibitory function aptamer-siRNA delivery system for HIV-1 therapy. *Mol Ther*, **16**. pp. 1481–1489.
79. Neff, CP, *et al.* (2011). An aptamer-siRNA chimera suppresses HIV-1 viral loads and protects from helper CD4(+) T cell decline in humanized mice. *Sci Transl Med*, **3**. pp. 66ra66.
80. Sattentau, QJ, and Moore, JP (1993). The role of CD4 in HIV binding and entry. *Philos Trans R Soc Lond B Biol Sci*, **342**. pp. 59–66.
81. Dalgleish, AG, Beverley, PC, Clapham, PR, Crawford, DH, Greaves, MF, and Weiss, RA (1984). The CD4 (T4) antigen is an essential component of the receptor for the AIDS retrovirus. *Nature*, **312**. pp. 763–767.
82. Wheeler, LA, *et al.* (2011). Inhibition of HIV transmission in human cervicovaginal explants and humanized mice using CD4 aptamer-siRNA chimeras. *J Clin Invest*, **121**. pp. 2401–2412.
83. Guo, P (2010). The emerging field of RNA nanotechnology. *Nat Nanotechnol*, **5**. pp. 833–842.
84. Shu, Y, Cinier, M, Shu, D, and Guo, P (2011). Assembly of multifunctional phi29 pRNA nanoparticles for specific delivery of siRNA and other therapeutics to targeted cells. *Methods*, **54**. pp. 204–214.
85. Guo, P (2005). RNA nanotechnology: engineering, assembly and applications in detection, gene delivery and therapy. *J Nanosci Nanotechnol*, **5**. pp. 1964–1982.
86. Farokhzad, OC, Jon, S, Khademhosseini, A, Tran, TN, Lavan, DA, and Langer, R (2004). Nanoparticle-aptamer bioconjugates: a new approach for targeting prostate cancer cells. *Cancer Res*, **64**. pp. 7668–7672.
87. Farokhzad, OC, *et al.* (2006). Targeted nanoparticle-aptamer bioconjugates for cancer chemotherapy *in vivo*. *Proc Natl Acad Sci USA*, **103**. pp. 6315–6320.
88. Dhar, S, Gu, FX, Langer, R, Farokhzad, OC, and Lippard, SJ (2008). Targeted delivery of cisplatin to prostate cancer cells by aptamer functionalized Pt(IV) prodrug-PLGA-PEG nanoparticles. *Proc Natl Acad Sci USA*, **105**. pp. 17356–17361.
89. Gu, F, *et al.* (2008). Precise engineering of targeted nanoparticles by using self-assembled biointegrated block copolymers. *Proc Natl Acad Sci USA*, **105**. pp. 2586–2591.

90. Wang, AZ, *et al.* (2008). Superparamagnetic iron oxide nanoparticle-aptamer bioconjugates for combined prostate cancer imaging and therapy. *ChemMedChem*, **3**. pp. 1311–1315.

91. Kim, E, *et al.* (2010). Prostate cancer cell death produced by the co-delivery of Bcl-xL shRNA and doxorubicin using an aptamer-conjugated polyplex. *Biomaterials*, **31**. pp. 4592–4599.

92. Morris, SW, *et al.* (1994). Fusion of a kinase gene, ALK, to a nucleolar protein gene, NPM, in non-Hodgkin's lymphoma. *Science*, **263**. pp. 1281–1284.

93. Duyster, J, Bai, RY, and Morris, SW (2001). Translocations involving anaplastic lymphoma kinase (ALK). *Oncogene*, **20**. pp. 5623–5637.

94. Zhang, P, *et al.* (2009). Using an RNA aptamer probe for flow cytometry detection of CD30-expressing lymphoma cells. *Lab Invest*, **89**. pp. 1423–1432.

95. Zhao, N, Bagaria, HG, Wong, MS, and Zu, Y (2011). A nanocomplex that is both tumor cell-selective and cancer gene-specific for anaplastic large cell lymphoma. *J Nanobiotechnol*, **9**. p. 2.

Chapter 6

Bioresponsive Nanoparticles for the Intracellular Delivery of RNAi Therapeutics

Kenneth Alan Howard

Department of Molecular Biology and Genetics, Interdisciplinary Nanoscience Center, University of Aarhus, 8000 Aarhus C, Denmark

kenh@inano.au.dk

The process of RNA interference (RNAi) by which cellular gene expression is controlled through the specific action of small interfering RNA (siRNA) is set to revolutionalise disease treatment. A greater understanding of the RNA pathway has resulted in an expanding repertoire of potential RNAi-based drugs whose action is dependent on intracellular target interaction. This work describes the use of polymeric nanoparticles used to overcome intracellular barriers and facilitate interaction of RNAi triggers with the RNAi machinery. Focus will be given to bioresponsive nanoparticles based on copolypeptides and hyperbranched polymers for controlling the intracellular spatial and temporal effects of synthetic microRNA and siRNA.

6.1 Introduction

The discovery that double stranded RNA mediates cellular gene silencing by the process of RNAi has paved the way towards a new

Nanotechnology for the Delivery of Therapeutic Nucleic Acids
Edited by Dan Peer

ISBN 978-981-4411-04-2 (Hardcover), 978-981-4411-05-9 (eBook)
www.panstanford.com

therapeutic approach. The process of post-transcriptional gene silencing occurs through mRNA engagement by an RNA-induced silencing complex (RISC) triggered by siRNA duplexes [1–4]. Messenger RNA specificity is facilitated by complementary base pairing of the guide strand retained after duplex disassembly and passenger strand removal during the recruitment process into RISC. The sequence specific nature of siRNA that allows for exclusive inhibition of target genes along with the capability for exogenous siRNA to enter the RNAi pathway presents "RNAi triggers" as highly attractive drug candidates.

The therapeutic effectiveness of siRNA is compromised by susceptibility to nuclease degradation, rapid renal clearance, and reduced cellular entry due to its polyanionic and macromolecular nature. Delivery is a key determinant in realizing the full clinical potential of RNAi-based therapeutics [5]. Non-viral delivery systems and chemical modifications have been developed to maximise payload at target sites [6–9]. Our expanding knowledge of the RNAi pathway has allowed identification of key molecules within the RNAi cascade that has broadened our selection of potential drug candidates. The action of these molecules is dependent on engagement with the RNAi machinery at specific intracellular sites; therefore, cellular entry and modulation of intracellular trafficking is of crucial importance.

Unique properties occurring at the nanoscale can be utilized to improve delivery of RNAi therapeutics. Polycation-based nanoparticles (termed polyelectrolyte complexes or polyplexes) formed by electrostatic self-assembly of siRNA with polycations offer flexible designs to fulfil both extracellular and intracellular delivery requirements [10, 11]. The inclusion of polymers, linkages, or functional groups that exhibit conformational change in response to biological stimuli is a strategy to control spatial and temporal delivery of siRNA. This work describes the application of these "bioresponsive" polyplexes for modulation of intracellular trafficking and gene silencing of different RNAi triggers.

6.2 Repertoire of Potential RNAi Therapeutics

An expanding panel of potential RNAi drugs has been generated through the greater understanding of the RNAi pathway. The discovery that exogenous long double stranded RNA could induce gene silencing supports an ancient defence mechanism against unwanted cytoplasmic double stranded RNA e.g. of viral origin [12]. A closely related class of endogenously expressed non-coding RNA termed microRNA (miRNA) that plays a key role in post-transcriptional control of cellular processes has been found [13–15]. The exogenous siRNA and endogenous miRNA pathway converge downstream through cytoplasmic RISC and DICER involvement. Understanding the RNAi pathway is pivotal to identify therapeutic substrates to harness RNAi effects and to design a delivery strategy that enables engagement within the RNAi pathway.

The conventional therapeutic approach is introduction of synthetic 21-22 nt siRNA into the cell cytoplasm to target RISC and mediate enzymatic cleavage of mRNA located at perinuclear sites. The upstream cytoplasmic endonuclease DICER facilitates cleavage of long duplexes into 21-22 nt siRNA in preparation for RISC processing [16]. This has resulted in the development of synthetic DICER-substrates 25-27 nt in length that show potent RNAi effect attributed to more effective RISC incorporation as part of the natural processing scheme [17, 18]. The drawback of non-specific induction of cellular innate immune responses exhibited by longer duplexes can be overcome by incorporation of chemical modification such as 2′MeO within the DICER-substrate siRNA [19, 20].

MicroRNAs are transcribed as hairpin structured primary RNA transcripts (pri-miRNAs) and processed in the nucleus by the RNase enzyme Drosha into shorter 50-70 nt precursor miRNAs (pre-miRNAs). The pre-miRNAs are transported across the nuclear pore complex into the cytoplasm by the enzyme Exportin-5 (XPO5) [21] where they are processed into mature miRNAs by DICER [22] that mediates potent gene silencing by translational repression or mRNA cleavage. Recent attention has been directed towards tumour suppressive miRNA and tumour inductive miRNA (termed oncomiRs) [23, 24]. Nuclear delivery of pri- or pre-miRNA transcripts or miRNA mimics introduced directly into the cytoplasm can be used to facilitate miRNA upregulation. Alternatively, cytoplasmic delivery of antisense oligonucleotides (termed Antagomirs) [25, 26] can be used to inhibit homologous miRNA. The selected strategy is dependent on the role of the miRNA in disease. In addition to the central role in controlling cellular processes, miRNAs are attractive RNAi therapeutics due to the improved potency of gene silencing [27] and avoidance of interferon activation as a result of endogenous miRNA processing [28].

Knockdown of nucleus-restricted transcripts [29, 30] through the involvement of DNA methylation [31, 32] has been reported that may lead to an alternative class of transcriptional gene silencing (TGS) therapeutics. In order to engage the nuclear RNAi pathways, pri-, pre-miRNA, and TGS therapeutics must be delivered across the nuclear membrane into the nucleus.

Plasmid [33] and viral-driven [34] intracellular expression of short hairpin siRNA (shRNA) is an alternative to using synthetic duplexes. However, despite the shorter knockdown duration observed with synthetic duplexes, they allow accurate dosage control and the ability to install stabilisation and reduce non-specific effects through chemical modification. RNAi-based therapies are, therefore, dependent on enabling technologies for cellular entry and intracellular trafficking to engage in either the nuclear or cytoplasmic RNAi machinery (e.g. Drosha, DICER, and RISC).

6.3 Nanoparticle-Based Delivery of RNAi Therapeutics

Unique properties at the nanoscale have fuelled the development of nanoparticles for delivery of RNAi therapeutics [11, 35]. These include the predisposition of nanoscale particles to circumvent the renal clearance mechanism yet permitting size-mediated transfer across vasculature endothelium into tissue and cellular entry by the process of endocytosis. Furthermore, they have a capacity to incorporate a high siRNA payload that is protected from nuclease degradation, in addition to a large surface to volume ratio that maximises surface functionalisation effects. On the flip-side endocytotic uptake results in endosomal capture and consequent lysosomal breakdown. Moreover, nanoparticle association requires release of the siRNA payload as a prelude to interactions with the RNAi machinery.

6.3.1 Polycation-Based Nanoparticles

A versatile class of nanoparticles termed polyplexes are those based on entropy-driven particle self-assembly between siRNA and polycations [10, 11]. The formation of submicron spherical particles (~50–300 nm) results from electrostatic interaction between the anionic phosphate backbone in the siRNA and polymer-bearing cationic amines (referred to as the N : P ratio). Synthetic (e.g. polyethylenimine (PEI) [36], poly(amido amine) [37]) and natural (e.g. chitosan [38–40], atelocollagen [41, 42]) polycations that can influence the biological activity of the nanoparticles have been used. For example, PEI exhibits endosomolytic properties that facilitate cytosolic delivery [43] whilst chitosan is mucoadehesive that potentiates delivery across mucosal surfaces [44]. The high N : P ratio (NP) and use of high molecular weight polymers normally a necessity for stable nanoparticles results in a net positive surface charge that facilitates cellular interactions and cell uptake. Inclusion of high molecular weight polymers shown to improve delivery, however, can result in insufficient release of the nucleic acid cargo [45] and toxicity due to cellular membrane disruption and intracellular interactions [46].

6.3.2 Bioresponsive Systems

The incorporation of functional components during polymer synthesis or by post-modification can be used to control nanoparticle stability, cellular uptake, intracellular trafficking, and polymer toxicity [11, 47]. Attention has been focused on inclusion of bioresponsive components and linkages whose activity is triggered by the intracellular conditions such as endosomal pH [48–50] or the cellular redox gradient [51–53] to facilitate, for example, cytosolic release of nanoparticle siRNA. These next-generation bioresponsive systems aim to fulfil the intracellular delivery requirements of RNAi triggers, namely cellular uptake, endosomal escape, targeting to specific subcellular compartments and siRNA release. Development of a non-toxic polyplex system capable of modulating cytoplasmic and nuclear trafficking and facilitating siRNA release should improve the therapeutic potential of RNAi therapeutics.

In the following sections, we will focus on two different bioresponsive nanoparticle delivery systems, the "reducible co-polypeptide (rCPP)" [54] and "reducible hyperbranched (rHB)" [55] systems, for different RNAi triggers developed in our laboratory at the Interdisciplinary Nanoscience Center (iNANO) at the University of Aarhus Denmark (in collaboration with the Oupicky Lab, Wayne State University, USA) (Fig. 6.1). The rational for their use is based on controlling (1) the intracellular trafficking required for RNAi triggers to engage at different levels of the RNAi cascade and (2) nanoparticle disassembly to potentiate incorporation into the RNAi machinery and tune biological activity. Incorporation of functional components at different molar ratios during the polymerisation process in these systems allows detailed investigations into the role and polymer content requirement for optimal intracellular spatial and temporal RNAi effects.

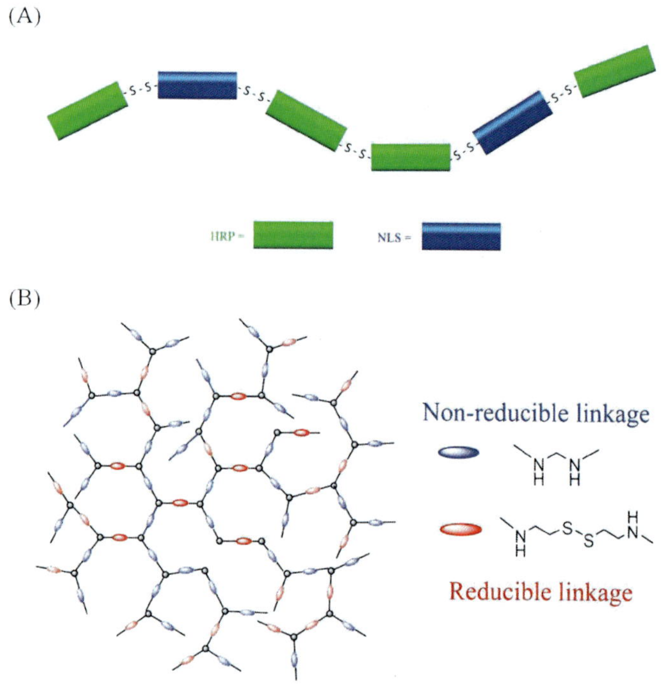

Figure 6.1 Schematic representation of bioresponsive polymers described in this chapter (A) multiblock reducible copolypeptide (rCPP) system (adapted from reference 56). (B) Reducible hyperbranched (rHB) system (adapted from reference 55). In both systems, the ratio of each component can be varied to study the structural–activity relationship.

6.4 Copolypeptide System

The bioreducible copolypeptide (rCPP) system is based on a series of polymers composed of different molar ratios of a lysine-containing histidine-rich peptide (HRP) CKHHHKHHHKC and SV40 Large T antigen-derived nuclear-localisation-sequence (NLS) peptide CGAGPKKKRKVC (Fig. 6.1). These polymers were prepared in the Oupicky laboratory by oxidative copolymerization of cysteine-capped HRP and NLS peptide subunits to form a linear copolypeptide backbone covalently linked by reducible disulphide bridges [56] (Fig. 6.1A). The multiblock strategy aims to install several properties into a single multifunctional system, namely particle formation, cellular uptake, cytosolic or nuclear localisation, and siRNA release. The HPR contains three cationic lysines that facilitate nanoparticle self-assembly and cellular binding via electrostatic interaction with phosphate-bearing siRNA and the cellular membrane. HRP contains histidyl groups that exhibit buffering capacity over the endosomal and lysosomal pH range and facilitate proton sponge-mediated endosomal escape and consequent siRNA cytosol localisation [57, 58]. Targeting to the nuclear compartment is achieved by inclusion of a SV-40 large T antigen nuclear localisation peptide sequence [59–61]. Polymer breakdown, subsequent nanoparticle disassembly, and siRNA release are triggered by the susceptibility of disulphide linkages to redox-induced cleavage inside the cell [62].

Synthesis of polymers with different molar ratio of HRP and NLS allows design selection for optimal siRNA delivery to cytoplasmic or nuclear targets. In a series of experiments, we investigated the intracellular delivery of nuclear (precursor miRNA transcripts and transcriptional silencing RNA) or cytoplasmic (21-mer siRNA) RNAi triggers [54]. Cellular tracking and RNAi activity were investigated in a range of polymers at different NLS: HRP ratio, rCPP-A (100 : 0), -B (75 : 25), -C (50 : 50), -D (25 : 75), -E (0 : 100). In order to study the intracellular effect of the HRP or NLS components, it was crucial to ensure polyplexes exhibited similar physiochemical properties such as hydrodynamic diameter and surface charge known to influence cellular interactions and uptake. Discrete particles in the size range 200–300 nm were formed at NP 10 with all the polymers by simple polymer addition to a 21-mer siRNA solution followed by gentle pipette mixing and incubation for ~1 h. The net positive surface charge ~20 mV that results from excess polymer is thought to contribute to cell binding with membrane-associated anionic proteins, sulphated polysaccarides or proteoglycans [63]. This allied with nanoscale particles with a predisposition for cellular endocytosis suggests a non-specific endocytosis process for the uptake observed by fluorescent microscopy in HeLa cells.

The incorporation of disulphide bridges within the polymer backbone is an established approach to facilitate intracellular polyplex disassembly triggered by the cleavage of disulphides by intracellular reducing agents such as glutathione [51, 52, 64–66]. Electrophoretic migration of siRNA in a polyacrylamide gel shift assay and morphological changes visualised by atomic force microscopy (AFM) were used to study particle decomplexation and subsequent siRNA release after addition of the reducing agent dithiothreitol (DTT) to mimic intracellular redox conditions. The requirement for redox conditions for particle disassembly and subsequent siRNA migration was shown for all reducible polymer systems at NP 5 and 10. Furthermore redox-induced nanoscale morphological changes were revealed by AFM. Interestingly, the average size of the particle increased from ~19 to ~32 nm after addition of DTT, which suggests a decomplexation process that includes particle swelling events possibly mediated by electrostatic repulsive forces between hydrated polymer chains during siRNA liberation. We have previously observed this phenomenon for redox-activated release of DNA using a similar polymer [67]. Cellular cytotoxity of polycations has been shown to be molecular weight dependent attributed to greater interaction of high charge density with cellular components. Intracellular breakdown into lower molecular weight fragments is a strategy to reduce polymer-associated intracellular toxicity [68]. No cytotoxity was observed for the reducible systems containing excess polymer (NP 20) in two cell lines, suggesting possible redox-induced polypeptide fragmentation into low molecular polymers and concomitant reduction in binding affinity to vital intracellular components.

Intracellular trafficking can be restricted by particle sequestration within the endosomal–lysosomal compartments as a consequence of uptake by endocytosis. Endosomal escape is, therefore, crucial and a prerequisite to enabling siRNA engagement with either cytosolic or nuclear targets. Movement from the endosomes into the cytoplasm and nuclear compartments were targeting considerations addressed with these systems. Live cellular uptake and intracellular trafficking were investigated in HeLa cells by fluorescent microscopy focused on particles composed of Alexa488-labelled rCPP-D (NLS 25 : 75 HRP) and Cy5-labelled siRNA. Fluorophore co-localisation at 2 h post-transfection suggests incomplete disassembly in a proportion of the particles, possibly reflecting sequestration within the endosomes and exposure within a non-reducing environment [69], although redox enzymes have been suggested to exist within endosomes and lysosomes [62]. Endosomal compartmentalisation was supported by co-localisation of a late endosomal plasmid-expressed marker cyan fluorescent (CFP)-TI-VAMP and directed particle movements suggestive of endosomal transportation along microtubule motor proteins [70]. Insufficient endosomal release could be attributed to insufficient histidyl content (six per peptide) within the polyplex or requirement for escape times greater than the 2 h investigated. Fluorescence resolution limits for identification of free siRNA in contrast to easily distinguishable punctuate fluorescence of nanoparticles could, however, lead to an underestimation of intracellular released siRNA. Furthermore, some polyplexes showed co-localisation with Hoechst-stained nuclei supporting some degree of endosomal escape and an effect of including an NLS peptide (25%). NLS-dependent nuclear localisation was demonstrated by confocal laser scanning microscopy 1 h post-transfection in fixed HeLa cells (Fig. 6.2). All polyplexes containing an NLS component showed some accumulation within the nuclei counter-stained with Hoechst except rCPP-A consisting of 100% NLS possibly reflecting particle instability or inadequate endosomal escape due to lack of HRP. In contrast, polyplexes devoid of NLS (rCPP-E) showed a perinuclear accumulation. These studies demonstrate the importance for a correct histidine/NLS balance for nanoparticle stability, endosomal escape, and nuclear trafficking.

The capability for rCPP to facilitate cytoplasmic and nuclear target engagement and mediate RNAi activity was demonstrated with conventional siRNA and transcriptional silencing siRNA. Delivery into cytoplasmic located RISC and consequent silencing with the rCPP series were evaluated using enhanced green fluorescent protein (EGFP)-specific siRNA in a H1299 endogenous-expressing EGFP cell line. Gradual increase in gene silencing, revealed by decreases in EGFP mean fluorescence intensity detected by flow cytometry, was observed from polymer A to E that correlates with increased HPR content. Nanoparticle formed at both NP 5 and 10 followed this trend with greatest gene silencing ~50% 48 h post-transfection shown for polymer E containing 100% HPR. The 50% gene silencing level may be due to insufficient endosomal escape; however, silencing indicates a degree of cytoplasmic localisation.

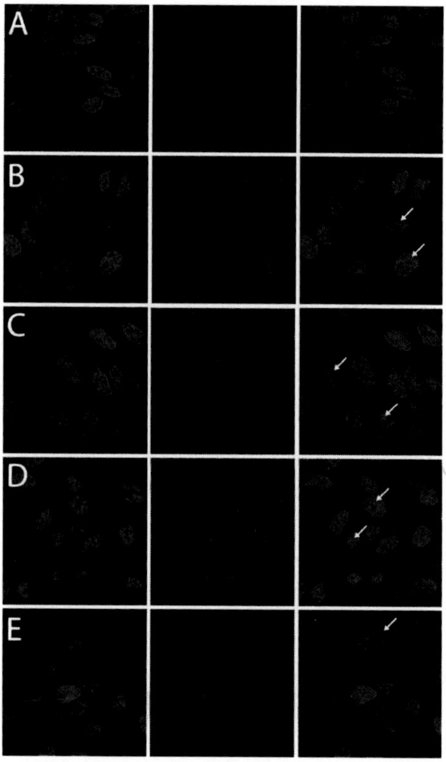

Figure 6.2 Nuclear localization of siRNA depends on the amount of NLS in the rCPP. HeLa cells were transfected for 1 h with 10 nM Cy5-labelled siRNA and polymers A–E (N : P 10), fixed and visualized using confocal laser scanning microscopy. (A) Polyplex A (100% NLS); (B) polyplex B (75% NLS); (C) polyplex C (50% NLS); (D) polyplex D (25% NLS); and (E) polyplex E (0% NLS). Left panels (DAPI stained nuclei, blue), centre panels (siRNA-Cy5, red), and right panels (overlay). Arrows in B, C, and D mark nuclear siRNA. The arrow in E marks perinuclear siRNA. From *J Gene Med* 2008; 10: 81–93.

Transcriptional-silencing siRNA targeting an essential promoter region of the elongation factor 1 alpha (EF1A) was used to investigate nuclear delivery in HeLa cells. Cytoplasmic EF1A mRNA levels were quantified by real-time RT-PCR at 48 h post-transfection with the rCPP series. EF1A levels standardised to GAPDH control showed a gradual decline with increased NLS content showing a clear NLS-dependency. Maximal reduction (~60%) was observed for polymer A composed of 100% NLS compared to minimal reduction (~10%) with polymer E without NLS. The capability for slight reductions without NLS could reflect cellular division during the 48 h period. The dependency of NLS inclusion for nuclear silencing indicates retained polymer association to the siRNA during the nuclear trafficking step. This suggests incomplete redox-mediated disassembly of the nanoparticle. High glutathione levels in the cytosol have been reported [62, 69]; however, nuclear-localised glutathione has been found [71]. Incomplete removal of the polymer may be exploited to ensure polymer-mediated nuclear trafficking prior to complete disassembly within the nucleus, enabling nuclear target interactions. It would be interesting to determine the amount retained or possible adverse effects on target engagement due to steric hindrance. This is an underexplored field; however, we have previously used restriction-site cleavage in plasmid DNA as a method to investigate the availability to specific interactions before and after redox-induced removal of polymer material [67].

Downregulation of miRNA associated with particular disease states and the potent RNAi effects observed with nuclear-processed primary miRNA transcripts (pri-miRNA) are the motivation for pri-miRNA delivery to the nuclear compartment. We investigated nuclear delivery and subsequent nuclear processing of pri-miRNA23a (183 nt) into mature miRNA23a (21 nt) effector molecules in the cytoplasm by inclusion of a NLS peptide derived from the importin α binding SV40 large T antigen. The RNA was isolated 48 h post-transfection in HeLa cells and the levels of mature RNA identified by northern analysis using an oligonucleotide probe complementary to miRNA23a. We found there was an NLS-dependent effect on the amount of mature miRNA detected. Increased levels of detected mature miRNA followed increases in the NLS content from rCPP-E (0% NLS) to rCPP-D (25% NLS) to rCPP-C (50% NLS), seemingly reflecting improved trafficking to the nucleus required for entry into the miRNA biogenesis pathway. Interestingly, further increase in NLS to 75% (rCPP-B) and 100% (rCPP-A) resulted in a decline in the amount of miRNA detected. This may reflect insufficient histidine content in these polymers, which limits endosomal release required for nuclear engagement. The low histidine content and concomitant reduced cationic lysines may also result in polyplex instability and consequent removal of the NLS polymer prior to nuclear engagement. The capability to alter the ratio of polymer constituents allows determination of the correct HPR/NLS balance required for successful nuclear pri-miRNA processing or cytoplasmic RISC-mediated siRNA silencing. Increased ratio of one component may, however, compromise an essential property provided by the other. These studies clearly highlight a structural–activity relationship whereby polymer composition determines intracellular compartmentalisation and consequent RNAi activity.

6.5 Hyperbranched System

This section describes the application of a bioreducible hyperbranched polymer for nanoparticle-based delivery of RNAi triggers [55]. This approach aims to modulate RNAi level and duration by tuning the release rate of siRNA by altering the reducible disulphide content within the polymer.

A series of rHB poly(amido amine) (PAMAM) polymers were synthetized in the Oupicky laboratory [72] by reaction of trifunctional amines to reducible bisacrylamide derivatives (Fig. 6.1B). Michael addition copolymerization was performed using equal molar ratio of triamine (1-(2-aminoethyl) piperazine) (AEPZ) and bisacrylamide monomers, *N,N*2-cystaminebisacrylamide (CBA), and *N,N*2-hexamethylenebisacrylamide (HMBA). Varying the ratio of the reducible CBA to non-reducible HMBA allows production of a series of polymers that contain different levels of disulphide bridges shown to change the susceptibility to redox conditions in this polymer [73]. A prominent feature of the hyperbranched design is the high charge density comprising of available free tertiary amines that facilitate polyelectrolyte complex formation with anionic siRNA and secondary and tertiary amines that allow endosomal escape by combined proton sponge [74] and polymer-induced pore formation [75] mechanisms. Extracellular nanoparticle stability is a prerequisite for siRNA protection for nanoparticle transit in the bloodstream and cellular uptake; however, intracellular disassembly is required to enable RNAi engagement. These requirements can be fulfilled by exploiting the redox potential gradient that exists between the extracellular and intracellular environment as a consequence of the high levels of reducing agents such as glutathione inside the cell compared to plasma [71] that can be used to cleave redox-sensitive disulphide bridges built into the bulk polymer. We investigated the relationship between disulphide content, particle disassembly, siRNA release, and RNAi activity.

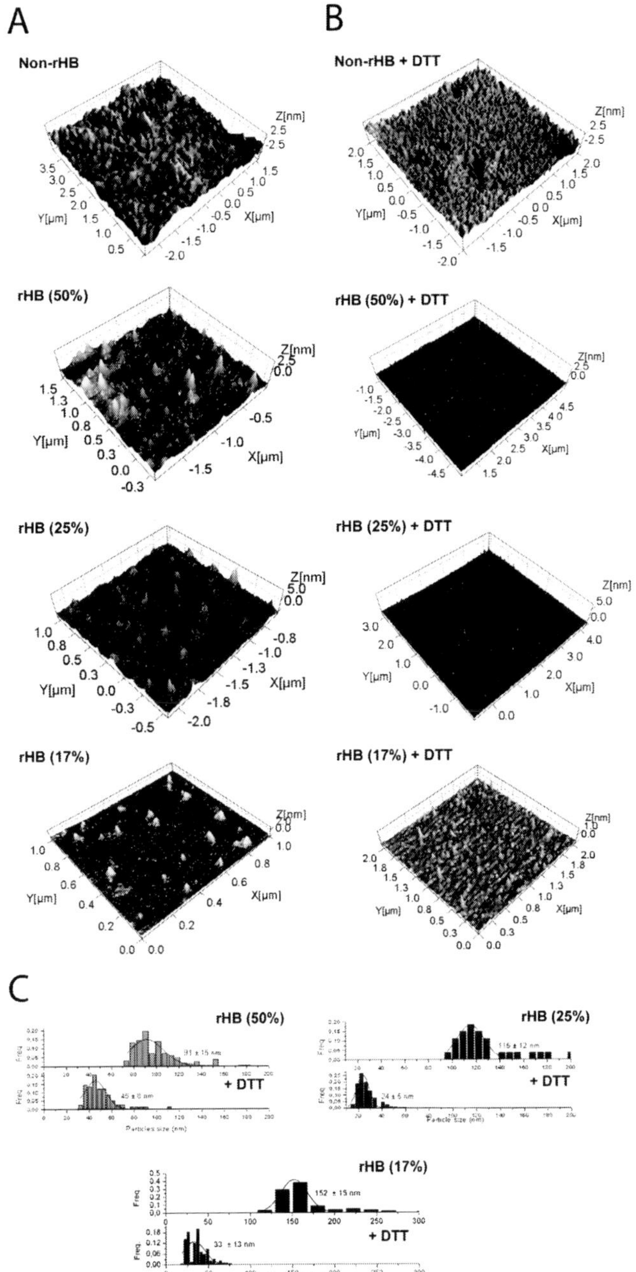

Figure 6.3 rHB sizes and reducibility. Atomic force microscopy images of polyplexes without (A) or with (B) incubation for 30 min in 25 mM dithiothreitol (DTT) at 37°C. (C) Particle size distribution with a best-fit curve superimposed for rHBs without DTT (top panels) and with DTT (lower panels). From *J Drug Target,* 2010; 18(10): 812–820.

The series of rHB polymers increasing in reducible disulphide content (rHB 17%, rHB 25%, rHB 50%) and control without disulphide (non-rHB) were used to form polyplexes with conventional siRNA by simple addition and gentle mixing at a weight: weight (w/w) ratio of 50:1 (polymer:siRNA). AFM studies revealed the formation of nanoscale systems ~100 nm for all the polymers (Fig. 6.3). In contrast to the non-reducible system, nanoparticle incubation with the reducing agent DTT resulted in morphological changes and accompanying size decreases for all the disulphide-containing polyplexes; rHB 17% from ~150 to ~33 nm, rHB 25% from ~110 to ~25 nm, and rHB 50% from ~90 to ~50 nm. Polyplexes formed with rHB 17% exhibited a greater net positive surface charge of ~30 mV compared to 11 mV and 9 mV for rHB 25% and 50%, respectively, most probably reflecting the higher molecular weight of this polymer (rHB 17% mw115,000; rHB 25% mw 32,000; rHB 50% mw 40,000) contributing to excess free charges after particle assembly. Even at w/w 50, all nanoparticle compositions exhibited non-toxicity in an EGFP-endogenous expressing H1299 cell line after 4 h incubation. This finding is similar to the non-toxicity previously shown for linear [76] and branched [73] reducible PAMAM polymers. Interestingly, the non-rHB also showed minimal toxicity in this cell line, suggesting the non-toxic effects of the hyperbranched PAMAM at the 50 kDa molecular weight used for this non-reducible polymer.

Cellular uptake studies performed in H1299 cells showed rapid internalisation for all the polyplexes containing Cy3-labelled siRNA over the 4 h period investigated; however, greater uptake was observed for rHB 17%, which could be related to the higher surface charge facilitating greater interaction with the cellular membrane. This was accompanied by diffuse fluorescence that indicates siRNA release. In contrast, punctuate fluorescence was observed for the non-rHB, rHB 25% and 50 %, which could indicate endosomal capture or a slower decomplexation rate with these polymers.

The capability of the rHB polymers for delivery of a range of RNAi triggers other than conventional siRNA was investigated with cytoplasmic delivery of pre-miRNA that engages upstream of RISC into DICER. The motivation for its use was driven by reports of improved silencing for endogenous sequential processing of miRNA [27], in addition to the potent effects reported for Dicer-substrate siRNA [17]. H1299 cells were transfected for 4 h with a series of rHBs (w/w 50) complexed with a 57 nucleotide pre-miRNA-23a. After 48 h, total RNA was isolated from the cells and northern analysis performed using a DNA oligonucleotide probe complementary to the miRNA-23a guide strand to quantify the delivery of pre-miRNA. Only the rHB (17%) polymer facilitated efficient transfection of pre-miRNA. This finding is consistent with the higher fluorescent uptake and higher knockdown efficiency observed using this polymer.

Knockdown experiments in H1299 EGFP-expressing cells, measured by the decrease in mean fluorescence detected by flow cytometry, was used to find optimal w/w of the range of polymers at 10, 20, or 40. In all HB polymers, higher w/w induced greater knockdown with the highest knockdown of ~74% (compared to non-treated control cells) observed for the rHB 17% w/w ratio 40. The necessity for excess polymer (high w/w ratio) indicates the key role of free polymer for particle stability, cellular uptake, and endosomal escape. The rHB with 25% reducible content showed the lowest EGFP knockdown of ~25%, which could reflect insufficient disassembly. Interestingly, the highest reducible content rHB 50% showed an intermediate level of knockdown ~45%, which suggests that other factors other than reducibility have an effect on silencing. Furthermore, the polyplexes composed of non-reducible polymer showed ~53% indicating some degree of disassembly to allow siRNA interaction in the absence of disulphides. The weak nature of electrostatic interactions constituting polyplex assembly may account for this effect. The work does, however, suggest level and duration of RNAi activity is dependent on the non-reducible/reducible balance.

The transient nature of synthetic siRNA requires persistent delivery of the siRNA into the cells for RNAi-based treatment of chronic diseases. Vector-based viral [34] or plasmid delivery [33] may overcome this requirement. An alternative approach is the application of sustained-release technology for controlling the siRNA release rate. Sustained-release microsphere [77] and nanoparticle [78] systems based on biodegradable poly(lactide-co-glycolide) (PLGA) have been developed for the delivery of siRNA. Approaches based on the degradation properties of PLGA, however, seem most suited to depot strategies rather than direct entry into cells.

We investigated the possibility to tailor the duration of RNAi by tuning the intracellular release of the siRNA controlled by disulphide content. We conducted experiments to investigate the level of disulphide content on temporal knockdown over a 72 h time course. The non-reducible polymer showed maximal EGFP silencing at 24 h, which decreased over the remaining 48 h period, which may reflect RNA degradation or dilution due to cellular division events. The reducible polyplexes followed a temporal incremental decrease suggestive of a sustained release effect. The rHB 50% showed incremental decreases at 24, 48, and 72 h and exhibited the most prominent change (25%) for the 24–72 h period, which could be due to a slower decomplexation process despite the highest disulphide content. The rHB 25% and rHB 17% showed a nominal decrease in EGFP silencing between 48 and 72 h, which was near to maximal at 24 h, suggesting faster decomplexation with the less redox-susceptible nanoparticle. The inherent instability of polyplexes due to low charge density contributed by short RNA adds to the difficulty in tuning the RNAi effects of system based on disulphide levels. Furthermore, the different cellular uptake and distribution patterns between the nanoparticles highlight the difficulty in reducing differences in RNAi effects solely to intracellular assembly and that other factors are involved in the multistep process of gene silencing. This approach, however, does suggest a role of disulphide content for modulation of the biological activity.

6.6 Conclusion

Delivery is central to realizing the clinical potential of RNAi therapeutics. Much attention has been focused on the extracellular delivery requirements of extended blood circulation, stability, and accumulation in target tissue. The RNAi process, however, depends on intracellular engagement, and so, intracellular delivery considerations of cellular uptake, cellular trafficking, and target interaction need to be addressed. The wide repertoire of potential RNAi therapeutics that engages at different levels of the RNAi cascade within different subcellular sites requires modulation of intracellular trafficking. Nanoparticle-based delivery is a promising approach to overcome both extracellular and intracellular barriers. A major challenge is to install extracellular stability without compromising intracellular disassembly and availability for siRNA target interaction. These requirements may be fulfilled by bioresponsive polycation-based nanoparticles whose action is triggered by local biological stimuli. The ability to incorporate functional components during polymer synthesis or by post-modification can be used to control cellular uptake, intracellular trafficking, and nanoparticle disassembly. Furthermore, these properties may offer the possibility to modulate the level and duration of RNAi activity. The structural–activity relationship exhibited by the polymers described in this work highlights the importance of polymer design for development of next-generation multifunctional delivery systems for RNAi therapeutics.

References

1. Fire A, Xu S, Montgomery MK, Kostas SA, Driver SE, and Mello CC. Potent and specific genetic interference by double-stranded RNA in Caenorhabditis elegans. *Nature*. 1998; **391**(6669):806–811.
2. Montgomery MK, Xu S, and Fire A. RNA as a target of double-stranded RNA-mediated genetic interference in Caenorhabditis elegans. *Proc Natl Acad Sci USA*. 1998; **95**(26):15502–15507.
3. Zamore PD, Tuschl T, Sharp PA, and Bartel DP. RNAi: double-stranded RNA directs the ATP-dependent cleavage of mRNA at 21 to 23 nucleotide intervals. *Cell*. 2000; **101**(1):25–33.
4. Elbashir SM, Harborth J, Lendeckel W, Yalcin A, Weber K, and Tuschl T. Duplexes of 21-nucleotide RNAs mediate RNA interference in cultured mammalian cells. *Nature*. 2001; **411**(6836):494–498.
5. Howard KA. Delivering the goods: realizing the clinical potential of RNAi. *Nanomedicine (Lond)*. 2009; **4**(6):595–598.
6. Davidson BL, and McCray PB, Jr. Current prospects for RNA interference-based therapies. *Nat Rev Genet*. 2011; **12**(5):329–340.
7. Rettig GR, and Behlke MA. Progress toward *in vivo* use of siRNAs-II. *Mol Ther*. 2012; **20**(3):483–512.
8. de Fougerolles A, Vornlocher HP, Maraganore J, and Lieberman J. Interfering with disease: a progress report on siRNA-based therapeutics. *Nat Rev Drug Discov*. 2007; **6**(6):443–453.
9. Davis ME, Zuckerman JE, Choi CH, Seligson D, Tolcher A, Alabi CA, *et al.* Evidence of RNAi in humans from systemically administered siRNA via targeted nanoparticles. *Nature*. 2010; **464**(7291):1067–1070.
10. Howard KA, and Kjems J. Polycation-based nanoparticle delivery for improved RNA interference therapeutics. *Expert Opin Biol Ther*. 2007; **7**(12):1811–1822.
11. Howard KA. Delivery of RNA interference therapeutics using polycation-based nanoparticles. *Adv Drug Deliv Rev*. 2009; **61**(9):710–720.

12. Cerutti H, and Casas-Mollano JA. On the origin and functions of RNA-mediated silencing: from protists to man. *Curr Genet.* 2006; **50**(2):81–99.

13. Lagos-Quintana M, Rauhut R, Lendeckel W, and Tuschl T. Identification of novel genes coding for small expressed RNAs. *Science.* 2001; **294**(5543):853–858.

14. He L, and Hannon GJ. MicroRNAs: small RNAs with a big role in gene regulation. *Nat Rev Genet.* 2004; **5**(7):522–531.

15. Ambros V. The functions of animal microRNAs. *Nature.* 2004; **431**(7006):350–355.

16. Bernstein E, Caudy AA, Hammond SM, and Hannon GJ. Role for a bidentate ribonuclease in the initiation step of RNA interference. *Nature.* 2001; **409**(6818):363–366.

17. Kim DH, Behlke MA, Rose SD, Chang MS, Choi S, and Rossi JJ. Synthetic dsRNA Dicer substrates enhance RNAi potency and efficacy. *Nat Biotechnol.* 2005; **23**(2):222–226.

18. Rose SD, Kim DH, Amarzguioui M, Heidel JD, Collingwood MA, Davis ME, *et al.* Functional polarity is introduced by Dicer processing of short substrate RNAs. *Nucleic Acids Res.* 2005; **33**(13):4140–4156.

19. Judge AD, Bola G, Lee AC, and MacLachlan I. Design of noninflammatory synthetic siRNA mediating potent gene silencing *in vivo*. *Mol Ther.* 2006; **13**(3):494–505.

20. Behlke MA. Chemical modification of siRNAs for *in vivo* use. *Oligonucleotides.* 2008; **18**(4):305–319.

21. Lund E, Guttinger S, Calado A, Dahlberg JE, and Kutay U. Nuclear export of microRNA precursors. *Science.* 2004; **303**(5654):95–98.

22. Lee Y, Jeon K, Lee JT, Kim S, and Kim VN. MicroRNA maturation: stepwise processing and subcellular localization. *EMBO J.* 2002; **21**(17):4663–4670.

23. Zhang B, Pan X, Cobb GP, and Anderson TA. microRNAs as oncogenes and tumor suppressors. *Dev Biol.* 2007; **302**(1):1–12.

24. Cho WC. OncomiRs: the discovery and progress of microRNAs in cancers. *Mol Cancer.* 2007; **6**:60.

25. Davis S, Lollo B, Freier S, and Esau C. Improved targeting of miRNA with antisense oligonucleotides. *Nucleic Acids Res.* 2006; **34**(8): 2294–2304.

26. Kurreck J. Antisense technologies. Improvement through novel chemical modifications. *Eur J Biochem.* 2003; **270**(8):1628–1644.

27. Boden D, Pusch O, Silbermann R, Lee F, Tucker L, and Ramratnam B. Enhanced gene silencing of HIV-1 specific siRNA using microRNA designed hairpins. *Nucleic Acids Res.* 2004; **32**(3):1154–1158.

28. Marques JT, Devosse T, Wang D, Zamanian-Daryoush M, Serbinowski P, Hartmann R, *et al.* A structural basis for discriminating between self and nonself double-stranded RNAs in mammalian cells. *Nat Biotechnol.* 2006; **24**(5):559–565.

29. Langlois MA, Boniface C, Wang G, Alluin J, Salvaterra PM, Puymirat J, *et al.* Cytoplasmic and nuclear retained DMPK mRNAs are targets for RNA interference in myotonic dystrophy cells. *J Biol Chem.* 2005; **280**(17):16949–16954.

30. Robb GB, Brown KM, Khurana J, and Rana TM. Specific and potent RNAi in the nucleus of human cells. *Nat Struct Mol Biol.* 2005; **12**(2):133–137.

31. Morris KV, Chan SW, Jacobsen SE, and Looney DJ. Small interfering RNA-induced transcriptional gene silencing in human cells. *Science.* 2004; **305**(5688):1289–1292.

32. Weinberg MS, Villeneuve LM, Ehsani A, Amarzguioui M, Aagaard L, Chen ZX, *et al.* The antisense strand of small interfering RNAs directs histone methylation and transcriptional gene silencing in human cells. *RNA.* 2006; **12**(2):256–262.

33. Brummelkamp TR, Bernards R, and Agami R. A system for stable expression of short interfering RNAs in mammalian cells. *Science.* 2002; **296**(5567):550–553.

34. Liu YP, Westerink JT, ter Brake O, and Berkhout B. RNAi-inducing lentiviral vectors for anti-HIV-1 gene therapy. *Methods Mol Biol.* 2011; **721**:293–311.

35. Singha K, Namgung R, and Kim WJ. Polymers in small-interfering RNA delivery. *Nucleic Acid Ther.* 2011; **21**(3):133–147.

36. Urban-Klein B, Werth S, Abuharbeid S, Czubayko F, and Aigner A. RNAi-mediated gene-targeting through systemic application of polyethylenimine (PEI)-complexed siRNA *in vivo. Gene Ther.* 2005; **12**(5):461–466.

37. Hoon Jeong J, Christensen LV, Yockman JW, Zhong Z, Engbersen JF, Jong Kim W, *et al.* Reducible poly(amido ethylenimine) directed to enhance RNA interference. *Biomaterials.* 2007; **28**(10):1912–1917.

38. Howard KA, Rahbek UL, Liu X, Damgaard CK, Glud SZ, Andersen MO, *et al.* RNA interference *in vitro* and *in vivo* using a novel chitosan/siRNA nanoparticle system. *Mol Ther.* 2006; **14**(4):476–484.

39. Liu X, Howard KA, Dong M, Andersen MO, Rahbek UL, Johnsen MG, *et al.* The influence of polymeric properties on chitosan/siRNA nanoparticle formulation and gene silencing. *Biomaterials.* 2007; **28**(6):1280–1288.

40. Howard KA, Paludan SR, Behlke MA, Besenbacher F, Deleuran B, and Kjems J. Chitosan/siRNA nanoparticle-mediated TNF-alpha knockdown in peritoneal macrophages for anti-inflammatory treatment in a murine arthritis model. *Mol Ther.* 2009; **17**(1):162–168.

41. Minakuchi Y, Takeshita F, Kosaka N, Sasaki H, Yamamoto Y, Kouno M, *et al.* Atelocollagen-mediated synthetic small interfering RNA delivery for effective gene silencing *in vitro* and *in vivo*. *Nucleic Acids Res.* 2004; **32**(13):e109.

42. Matoba T, Orba Y, Suzuki T, Makino Y, Shichinohe H, Kuroda S, *et al.* An siRNA against JC virus (JCV) agnoprotein inhibits JCV infection in JCV-producing cells inoculated in nude mice. *Neuropathology.* 2008; **28**(3):286–294.

43. Boussif O, Lezoualc'h F, Zanta MA, Mergny MD, Scherman D, Demeneix B, *et al.* A versatile vector for gene and oligonucleotide transfer into cells in culture and *in vivo*: polyethylenimine. *Proc Natl Acad Sci USA.* 1995; **92**(16):7297–7301.

44. Soane RJ, Frier M, Perkins AC, Jones NS, Davis SS, and Illum L. Evaluation of the clearance characteristics of bioadhesive systems in humans. *Int J Pharm.* 1999; **178**(1):55–65.

45. Schaffer DV, Fidelman NA, Dan N, and Lauffenburger DA. Vector unpacking as a potential barrier for receptor-mediated polyplex gene delivery. *Biotechnol Bioeng.* 2000; **67**(5):598–606.

46. Moghimi SM, Symonds P, Murray JC, Hunter AC, Debska G, and Szewczyk A. A two-stage poly(ethylenimine)-mediated cytotoxicity: implications for gene transfer/therapy. *Mol Ther.* 2005; **11**(6):990–995.

47. Breunig M, Bauer S, and Goepferich A. Polymers and nanoparticles: intelligent tools for intracellular targeting? *Eur J Pharm Biopharm.* 2008; **68**(1):112–128.

48. Shim MS, and Kwon YJ. Acid-responsive linear polyethylenimine for efficient, specific, and biocompatible siRNA delivery. *Bioconjug Chem.* 2009; **20**(3):488–499.

49. Convertine AJ, Benoit DS, Duvall CL, Hoffman AS, and Stayton PS. Development of a novel endosomolytic diblock copolymer for siRNA delivery. *J Control Release.* 2009; **133**(3):221–229.

50. Henry SM, El-Sayed ME, Pirie CM, Hoffman AS, and Stayton PS. pH-responsive poly(styrene-alt-maleic anhydride) alkylamide copolymers for intracellular drug delivery. *Biomacromolecules*. 2006; **7**(8):2407–2414.

51. Read ML, Singh S, Ahmed Z, Stevenson M, Briggs SS, Oupicky D, *et al.* A versatile reducible polycation-based system for efficient delivery of a broad range of nucleic acids. *Nucleic Acids Res*. 2005; **33**(9):e86.

52. Manickam DS, Li J, Putt DA, Zhou QH, Wu C, Lash LH, *et al.* Effect of innate glutathione levels on activity of redox-responsive gene delivery vectors. *J Control Release*. 2010; **141**(1):77–84.

53. Soundara Manickam D, and Oupicky D. Polyplex gene delivery modulated by redox potential gradients. *J Drug Target*. 2006; **14**(8): 519–526.

54. Rahbek UL, Howard KA, Oupicky D, Manickam DS, Dong M, Nielsen AF, *et al.* Intracellular siRNA and precursor miRNA trafficking using bioresponsive copolypeptides. *J Gene Med*. 2008; **10**(1):81–93.

55. Rahbek UL, Nielsen AF, Dong M, You Y, Chauchereau A, Oupicky D, *et al.* Bioresponsive hyperbranched polymers for siRNA and miRNA delivery. *J Drug Target*. 2010; **18**(10):812–820.

56. Manickam DS, and Oupicky D. Multiblock reducible copolypeptides containing histidine-rich and nuclear localization sequences for gene delivery. *Bioconjug Chem*. 2006; **17**(6):1395–1403.

57. Pichon C, Goncalves C, and Midoux P. Histidine-rich peptides and polymers for nucleic acids delivery. *Adv Drug Deliv Rev*. 2001; **53**(1):75–94.

58. Goncalves C, Pichon C, Guerin B, and Midoux P. Intracellular processing and stability of DNA complexed with histidylated polylysine conjugates. *J Gene Med*. 2002; **4**(3):271–281.

59. Bremner KH, Seymour LW, and Pouton CW. Harnessing nuclear localization pathways for transgene delivery. *Curr Opin Mol Ther*. 2001; **3**(2):170–177.

60. Zanta MA, Belguise-Valladier P, and Behr JP. Gene delivery: a single nuclear localization signal peptide is sufficient to carry DNA to the cell nucleus. *Proc Natl Acad Sci USA*. 1999; **96**(1):91–96.

61. Sebestyen MG, Ludtke JJ, Bassik MC, Zhang G, Budker V, Lukhtanov EA, *et al.* DNA vector chemistry: the covalent attachment of signal peptides to plasmid DNA. *Nat Biotechnol*. 1998; **16**(1):80–85.

62. Saito G, Swanson JA, and Lee KD. Drug delivery strategy utilizing conjugation via reversible disulfide linkages: role and site of cellular reducing activities. *Adv Drug Deliv Rev.* 2003; **55**(2):199–215.

63. Mislick KA, and Baldeschwieler JD. Evidence for the role of proteoglycans in cation-mediated gene transfer. *Proc Natl Acad Sci USA.* 1996; **93**(22):12349–12354.

64. Read ML, Bremner KH, Oupicky D, Green NK, Searle PF, and Seymour LW. Vectors based on reducible polycations facilitate intracellular release of nucleic acids. *J Gene Med.* 2003; **5**(3):232–245.

65. Read ML, Logan A, and Seymour LW. Barriers to gene delivery using synthetic vectors. *Adv Genet.* 2005; **53**PA:19–46.

66. Son S, Namgung R, Kim J, Singha K, and Kim WJ. Bioreducible polymers for gene silencing and delivery. *Acc Chem Res.* 2012; **45**(7):1100–1101.

67. Howard KA, Dong M, Oupicky D, Bisht HS, Buss C, Besenbacher F, *et al.* Nanocarrier stimuli-activated gene delivery. *Small.* 2007; **3**(1):54–57.

68. Breunig M, Lungwitz U, Liebl R, and Goepferich A. Breaking up the correlation between efficacy and toxicity for nonviral gene delivery. *Proc Natl Acad Sci USA.* 2007; **104**(36):14454–14459.

69. Austin CD, Wen X, Gazzard L, Nelson C, Scheller RH, and Scales SJ. Oxidizing potential of endosomes and lysosomes limits intracellular cleavage of disulfide-based antibody-drug conjugates. *Proc Natl Acad Sci USA.* 2005; **102**(50):17987–17992.

70. Murray JW, and Wolkoff AW. Roles of the cytoskeleton and motor proteins in endocytic sorting. *Adv Drug Deliv Rev.* 2003; **55**(11): 1385–1403.

71. Bellomo G, Vairetti M, Stivala L, Mirabelli F, Richelmi P, and Orrenius S. Demonstration of nuclear compartmentalization of glutathione in hepatocytes. *Proc Natl Acad Sci USA.* 1992; **89**(10):4412–4416.

72. Blacklock J, You YZ, Zhou QH, Mao G, and Oupicky D. Gene delivery *in vitro* and *in vivo* from bioreducible multilayered polyelectrolyte films of plasmid DNA. *Biomaterials.* 2009; **30**(5):939–950.

73. Chen J, Wu C, and Oupicky D. Bioreducible hyperbranched poly (amido amine)s for gene delivery. *Biomacromolecules.* 2009; **10**(10): 2921–2927.

74. Pack DW, Hoffman AS, Pun S, and Stayton PS. Design and development of polymers for gene delivery. *Nat Rev Drug Discov.* 2005; **4**(7):581–593.

75. Zhang ZY, and Smith BD. High-generation polycationic dendrimers are unusually effective at disrupting anionic vesicles: membrane bending model. *Bioconjug Chem.* 2000; **11**(6):805–814.

76. Lin C, Zhong Z, Lok MC, Jiang X, Hennink WE, Feijen J, *et al.* Novel bioreducible poly(amido amine)s for highly efficient gene delivery. *Bioconjug Chem.* 2007; **18**(1):138–145.

77. Khan A, Benboubetra M, Sayyed PZ, Ng KW, Fox S, Beck G, *et al.* Sustained polymeric delivery of gene silencing antisense ODNs, siRNA, DNAzymes and ribozymes: *in vitro* and *in vivo* studies. *J Drug Target.* 2004; **12**(6):393–404.

78. Hasan W, Chu K, Gullapalli A, Dunn SS, Enlow EM, Luft JC, *et al.* Delivery of multiple siRNAs using lipid-coated PLGA nanoparticles for treatment of prostate cancer. *Nano Lett.* 2012; **12**(1):287–292.

Chapter 7

Lipid-Like Delivery Materials for Efficient siRNA Delivery

James Dahlman,[a,b] Robert Langer,[a,b] and Michael Goldberg[c,*]

[a]*Harvard-MIT Division of Health Science and Technology, Cambridge, MA, 02139, USA*

[b]*David H. Koch Center for Integrative Cancer Research, MIT, Cambridge, MA, 02139, USA*

[c]*Dana-Farber Cancer Institute, Boston, MA, 02215, USA*

*Michael_Goldberg1@dfci.harvard.edu

Realizing the full potential of RNAi therapeutics will require the advent of safe and efficient carriers. Since the structure of the ideal vehicle remains unknown, the ability to generate and evaluate many compounds in a high-throughput manner is likely to direct our efforts. One-step conjugation chemistries facilitate the synthesis of large libraries, whose members can be screened for efficacy both *in vitro* and *in vivo*. Comparison of the top-performing compounds can yield structure–function relationships that can, in turn, inform the synthesis of iterative libraries. This chapter describes the synthesis, screening, formulation, evolution, and application of "lipidoids", a novel class of lipid-like molecules that highlights the utility of combinatorial approaches for the production of effective siRNA delivery vehicles.

Nanotechnology for the Delivery of Therapeutic Nucleic Acids
Edited by Dan Peer

ISBN 978-981-4411-04-2 (Hardcover), 978-981-4411-05-9 (eBook)
www.panstanford.com

7.1 Introduction

Before the discovery that RNA could regulate gene expression via RNA interference (RNAi) in 1998,[1] RNA was viewed primarily as an intermediary between DNA and protein. Since this breakthrough, the role of RNA in biology and medicine has been greatly expanded. RNA has been applied not only as a tool in basic research but also as a clinical therapeutic modality. The utility of RNA in these settings requires successful delivery of RNA into the cytoplasm of the cells of interest. *In vitro*, delivery necessitates that RNA pass through the anionic and amphiphilic cell membrane so that it can interact with the cytosolic RNA induced silencing complex (RISC) to initiate gene silencing. The physical properties of RNA render this feat challenging; specifically, RNA is large (~13 kDa for a double-stranded 21-mer) and negatively charged. *In vivo*, successful delivery is even more difficult. Before entering the cytoplasm, molecules must interact with the correct cell type while avoiding degradative nucleases and phagocytotic cells in the bloodsteam[2] (Fig. 7.1) . To overcome these obstacles, scientists have designed synthetic delivery vehicles from gold, cholesterol, polymers, and many other materials.[3] One subset that has delivered siRNA to cells efficiently *in vitro* and *in vivo* is a class of amphiphilic lipid-like structures, termed "lipidoids."

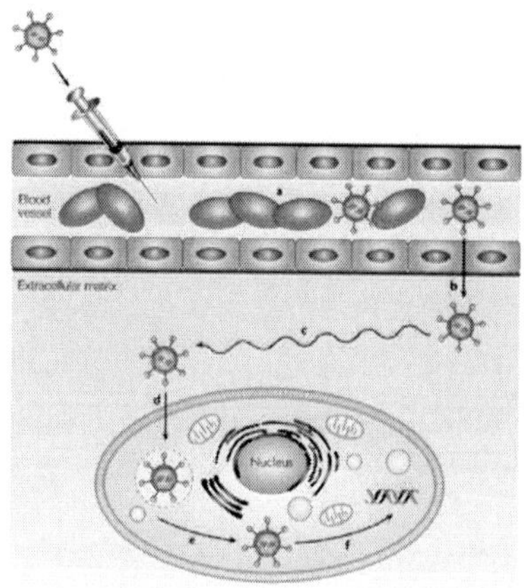

Figure 7.1 Physiological barriers to small RNA delivery. *In vivo* delivery requires that vehicles ferry RNA to the correct cell type while avoiding phagocytotic uptake, degradative nucleases, and inadvertent immune stimulation. If the carrier successfully completes this task, it must then transport the large, anionic nucleic acids across the negative lipid bilayer. Reproduced with permission from Ref. [2].

7.2 Motivation: Need for Novel siRNA Carriers *in vivo*

Originally reported in 2008, lipidoids were designed to deliver siRNA efficiently to hepatocytes. Cholesterol-conjugated siRNA was used to enable the first illustration of therapeutic silencing of an endogenous gene by systemic siRNA administration.[4] Lipids can facilitate siRNA transfection not only through direct conjugation to the payload but also through physical entrapment.[5,6] Though these early successes hinted at what types of materials might effectively deliver siRNA, the investigation of new materials was limited by difficult chemical syntheses, requiring multiple steps and purifications. The engineering community had faced similar issues related to the delivery of DNA, which shares many of the physical properties of RNA.

7.3 Approach: Efficient Chemistry Allows for High-Throughput Combinatorial Library Synthesis and Screening

To overcome these difficulties, a simple hypothesis — that biodegradable, cationic compounds might facilitate safe delivery — was used as rationale for the synthesis of a structurally diverse library of poly(β-amino esters).[7] To produce a large library without time-consuming protection and purification steps, diacrylates were reacted with amines via the Michael addition. This efficient scheme produced more than 2,000 distinct compounds, enough to justify relationships between chemical structure, particle size, and gene transfection efficiency.[7] The structure–function relationships from this library were used to design a second library,[8] which in turn led to the identification of a potent compound that was fine-tuned to optimize delivery.[9] These studies illustrated that a broad structural class of molecules could be screened to enable the isolation of a specific, robust compound through iterative libraries derived from efficient chemistry.[7–9] These large polymers, which were designed to bind long nucleic acids through electrostatic interactions, did not efficiently condense small RNAs, so an alternative structure was required.

Building on the success of iterative rational libraries, we investigated whether smaller amphiphilic molecules would effectively deliver small RNAs. To this end, cationic lipid-like structures were chosen for three reasons. First, cationic lipids had previously been shown to deliver siRNA.[5] Second, we hypothesized that smaller cationic compounds would bind to highly anionic siRNA tightly enough to prevent dissociation in the bloodstream, but loosely enough to release the RNA once it was inside the cytoplasm. Finally, we proposed that lipid tails — which facilitate interaction with membranes, enhancing uptake by the plasma membrane and destabilizing endosomal membranes[10] — might work in concert with amines — which condense nucleic acids and induce endosomal rupture through the proton sponge mechanism.[11] The pKa of amines allows them to absorb protons without reducing endosomal pH. The proton influx produces an ionic and osmotic gradient that causes water to flow into the endosome, which eventually bursts.[12]

To test this structural space, alkyl-acrylamide or alkyl-acrylate lipid tails were conjugated to amine backbones via the Michael addition, generating more than 700 first-generation lipidoids[13] (Fig. 7.2). Four structural parameters were modulated. First, alkyl-acrylamide and alkyl-acrylate tail length was varied between 9 and 18 fully saturated carbons. Second, sundry amine backbones were chosen in order to maximize structural diversity, thus increasing the probability of identifying critical structural motifs; the chemical space explored was, however, focused by knowledge obtained from analysis of previous libraries. Third, the bond linking the amine backbone to the alkyl tail was comprised of either a biodegradable ester or a stable amide. Finally, the impact of backbone charge was investigated by quaternizing amines with methyl iodide.[11] This alkylating agent imparted a permanent positive charge on the amine.

Once the library was synthesized, an efficient *in vitro* assay was required to screen the large number of structures. An ideal assay would measure target and control gene expression concurrently, so that potent silencing would not be confused with cytotoxicity. To achieve this, human cervical cancer (HeLa) cells were made to express stably both firefly and renilla luciferases (termed dual HeLa cells). Lipidoids were complexed with firefly luciferase-targeting siRNA (siFire) before the activities of both luciferases — which have different substrates — were measured by luminescence the following day. Measuring target and control gene expression concurrently with luminescence is efficient, permitting the rapid identification of effective compounds and structure–function relationships. In this case, lipidoids with at least two amines per backbone, two long lipid tails, or many shorter lipid tails potently silenced the target gene without influencing control gene expression.

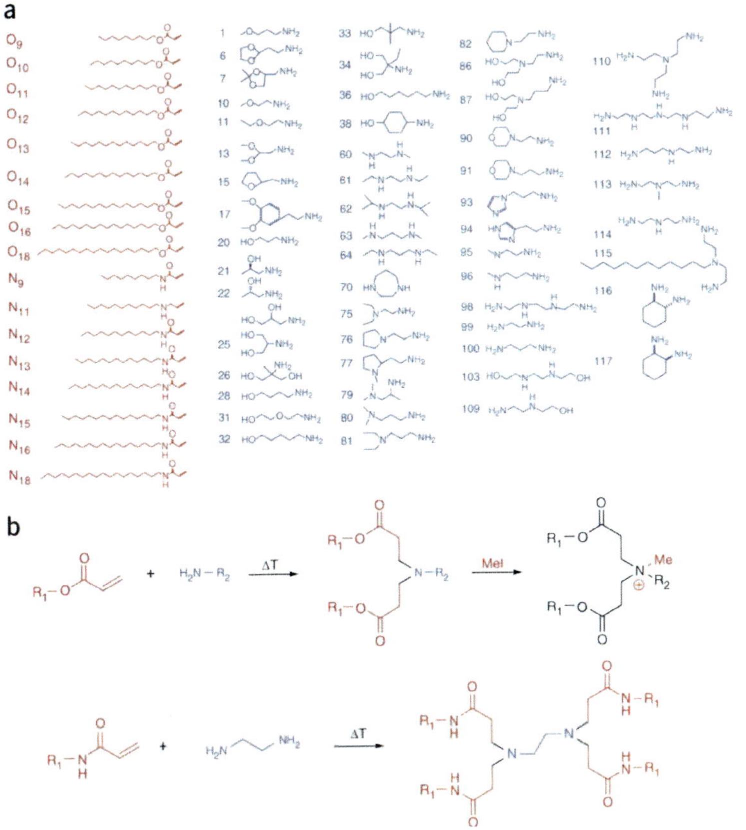

Figure 7.2 Synthesis of lipidoids. (a) Alkyl-acrylate, alkyl-acrylamide, and amino molecules were used to synthesize a combinatorial library of lipidoids. (b) Synthesis occurs through the conjugate addition of amines to an acrylate or acrylamide. Depending on the number of addition sites in the amino monomer, lipidoids can be formed with anywhere from 1 to 7 tails. Amino groups in the lipidoid can be quaternized by treatment with methyl iodide. For ease of nomenclature, lipidoids are named as follows: (amine number)(acrylate or acrylamide name)-(number of tails)("+" if quaternized). Reproduced with permission from Ref. [13].

These structure–function relationships were used to inform the synthesis of a second-generation lipidoid library, consisting of more than 500 new molecules. Of the 1,200 total structures, ~5% silenced target gene expression at least as efficiently as the positive control, Lipofectamine 2000, a commercially-available transfection reagent. The second-generation library led to the insights that compounds with amide linkages, more than two alkyl tails between 8 and 12 carbons, and one alkyl tail less than saturating substitution delivered siRNA to dual HeLa cells most efficiently. After identifying effective candidates in HeLa cells, the ability of compounds to silence genes in human hepatocellular carcinoma (HepG2) cells and primary bone marrow-derived murine macrophages was investigated. These additional *in vitro* assays were performed because transfection efficiencies can vary dramatically between different cell lines. Lipidoids were less proficient than Lipofectamine 2000 when silencing endogenous genes in HepG2 cells but more efficient when silencing endogenous genes in macrophages, with 50% target gene reduction at siRNA doses as low as 1 nM. By contrast, Lipofectamine did not silence macrophages even at 10 nM, suggesting lipidoids might be useful for transfecting cell lines that were previously refractory to treatment.[13]

7.4 Translation: Moving from *in vitro* to *in vivo* Screening

Based on the results of *in vitro* screens in multiple cell lines, 17 compounds were chosen for *in vivo* evaluation. Although *in vitro* results do not exactly correlate to *in vivo* results, they do identify a subset of materials that should be tested in animals, thereby saving both time and money. A desirable *in vivo* screen will measure gene expression of a target produced specifically by one cell type quickly and inexpensively. To this end, we measured hepatocyte delivery by targeting Factor 7 (F7). Hepatocytes were chosen as a target cell because they play an important role in cancer, hepatitis, hypercholesterolemia, malaria, and many other pathologies.[14] F7 was selected as a target gene, since it is found in plasma and has a short half-life. This means F7 levels can be measured in blood serum without sacrificing the mouse, and knockdown at the protein level can be quantified within hours of injection. Moreover, since F7 is produced specifically by hepatocytes, delivery is not confused with that to kupffer cells, immune cells that reside in the liver.[13]

Lipidoids were mixed with cholesterol and PEG-lipid conjugates prior to intravenous injection. These excipients had been shown to increase serum stability, resistance to aggregation, and *in vivo* tolerability.[15, 16] The molar ratios of the three *in vivo* ingredients was also shown to influence biophysical characteristics, *in vivo* potency, and tolerability;[17] more details relating to the importance of formulation will be presented later in this chapter. After administering a dose of 5 mg/kg, 5 of the 17 compounds silenced target protein expression by over 50%, with one (termed $98N_{12}$) silencing F7 by more than 90% relative to saline control. Before continuing with additional *in vivo* models, we discerned the active isomer of $98N_{12}$, whose reaction yields multiple products. The chemical mixture was separated into constituents that were distinguished by the number of tails attached to the amine backbone. A structure with five tails ($98N_{12}$-5), one less than saturating substitution, was identified as the active component, while the other variants were found to be relatively ineffective.

Purified $98N_{12}$-5 was tested to determine how much siRNA was required to silence a target gene by 50%, how long silencing persisted, whether F7 silencing resulted in the expected physiological effect, and whether it could mediate the silencing of another liver-specific gene, Apolipoprotein B (ApoB). $98N_{12}$-5 was shown to result in dose-dependent silencing that lasted more than two weeks. The silencing of F7 and ApoB led to physiologically relevant outcomes, namely increased clotting time and decreased LDL levels. Finally, the compound was evaluated in rats, which are used for pre-clinical immunological assays, and in non-human primates. It potently silenced endogenous target genes in both species.

$98N_{12}$-5 was selected from a library that varied the length of a fully saturated alkyl tail attached to amine backbones.[13] Since alkyl tail length was found to influence delivery dramatically, we explored whether other structural changes to the hydrophobic tails might impact efficacy.[18] Amine backbones 98 and 100 from the original lipidoid library were conjugated to 17 acrylates with oxygens, branched carbons, or amines in the tail. Experiments using dual HeLa cells confirmed that *in vitro* silencing was affected by tail structure. Conjugating PEG (CH_2CH_2O) directly to the amine backbone inhibited silencing, while attaching PEG-like repeats (for example, $CH_2CH(CH_3)O$ or $CH_2CH_2CH_2O$) resulted in compounds that were toxic *in vitro*. Contrasting this was potent, non-toxic silencing when a tail consisting of PEG monomer, an amine, and an acrylate was added. Although the precise mechanism governing tolerability was not identified, these results demonstrate that small changes in chemical structure can dramatically influence function. Despite promising results *in vitro*, new structures were outperformed by $98N_{12}$-5 *in vivo*, which reduced F7 expression by nearly 90% at 2 mg/kg.[18]

7.5 Optimization: Formulation Parameters Greatly Influence Carrier Efficacy

During the original investigation of $98N_{12}$-5, delivery was enhanced by purifying a chemical mixture into constituents differentiated by the number of tails attached to the amine backbone.[13] However, this chemical optimization neglects the effect of both PEG-lipid and cholesterol on delivery. Since these components are known to influence the stability, half-life, and distribution of nanoparticles, we investigated how the lipidoid:siRNA mass ratio, type of PEG-lipid, and particle size influenced delivery.[17]

The first variable modified was the lipidoid:siRNA mass ratio. High mass ratios increase the amount of lipid injected into the animal, which could cause toxic effects. Conversely, lowering the mass ratio could reduce the amount of siRNA entrapped by the particle, since fewer positive amines would be available to interact with the anionic RNA backbone. Non-entrapped siRNA would remain free in solution, effectively wasted. Consequently, siRNA entrapment was measured at mass ratios between 5 and 30. Entrapment was almost 100% at mass ratios between 10 and 30 but decreased dramatically at a mass ratio equal to 5. Consequently, efficacy and toxicity were tested at ratios of 7.5, 10, 15, 20, and 30. At a mass ratio of 7.5, compounds were well tolerated at all doses, while larger ratios resulted in weight loss at 10 mg/kg. Since efficacy decreased modestly, a mass ratio of 7.5 was selected for additional studies.

After optimizing the lipidoid:siRNA mass ratio, efficacy and tolerability as a function of the PEG-lipid tail length was evaluated at 2.5 and 25 mg/kg. PEG-lipids are anchored into the nanoparticle membrane via hydrophobic interactions. Consequently, adjusting the PEG tail length influences how securely PEG inserts into the particle. PEG-lipid length was varied between 10 and 16 carbons. At 2.5 mg/kg, mouse weight loss remained constant as lipid length changed, but differed dramatically at 25 mg/kg. At 25 mg/kg, lipids with 14, 15, or 16 carbons were tolerated better than chains between 10 and 13 carbons. At the same time, efficacy was maximized for tails 13 to 15 carbons in length. These studies demonstrated that alkyl tails with 14 or 15 carbons were optimal. Since even-numbered carbon tails are less expensive, C14 was chosen for subsequent studies, which sought to correlate particle size and efficacy.

To control particle size, $98N_{12}$-5 particles were made with identical molar ratios and extruded through membranes with different pore sizes. This produced particles with diameters of 150, 85, 60, and 50 nm, respectively. F7 levels measured 48 h after intravenous injection at a siRNA dose of 3 mg/kg clearly showed improved silencing as size decreased. These results were expected, since smaller particles enter into liver parenchyma through fenestrated hepatic vasculature more easily. Liver distribution was later verified using radioactively labeled $98N_{12}$-5. Following intravenous injection, particles were observed predominantly in the liver (92% of injected dose), with a small percentage found in the spleen, another fenestrated organ.

7.6 Synergy: Combining Existing Compounds to Achieve Improved Delivery

The iterative library approach identified $98N_{12}$-5 as the most potent molecule from an original library of over 1,200 compounds. A large percentage from the original library was ineffective, presumably because delivery requires both that the vehicles enter the endosome and escape it to enter the cytoplasm. Members of our lab reasoned that some lipidoids might promote endosomal uptake without affecting escape, and vice versa. They investigated whether combining two ineffective lipidoids would produce a synergistic effect on delivery.[19] They envisioned that one lipidoid would facilitate uptake while the second would foster escape.

To test this idea, 36 lipidoids were combined to form 630 binary pairs. The mass ratio of each lipidoid pair was varied between zero and one with a step size of 0.2, resulting in nearly 4,000 formulations.[19] Since the purpose of the library was to investigate the effects of binary lipidoid combinations, *in vitro* knockdown in dual

HeLa cells was analyzed mathematically by a term that quantified the difference between silencing with individual components and their binary combination, the *S* value. This value was defined such that complete silencing achieved with a mixture of lipidoids that did not silence on their own equaled 1, no difference between binary and individual lipidoids equaled 0, and no silencing from binary mixtures of lipidoids that completely silenced targets individually equaled −1. Although many combinations did not improve targeting ($S = 0$), approximately 5% of the tested compounds had *S* values greater than 0.5, indicating synergistic delivery. Several combinations were more effective *in vitro* than $98N_{12}$-5, silencing target gene expression more than 75% at 20 nM. Binary potency often increased as the first lipidoid:second lipidoid mass ratio approached 1:1 (the most evenly mixed combination). Synergistic delivery was also seen *in vivo*. Two components that silenced 20% when used independently reduced F7 expression by 85% when combined at a relative mass ratio of 3:1.

7.7 Next-Generation: Identifying Improved Carriers Using Innovative Chemistry

Through a series of iterative studies, first-generation lipidoids were optimized so that 50% reduction of a hepatic target gene was achieved when siRNA was injected intravenously at 1 mg/kg. Lipidoid synthesis revealed that new lipid delivery agents could be produced without time-consuming and expensive chemical synthesis. Utilizing three times as much lipid carrier, stable nucleic acid lipid particles (SNALP) formulations were also shown to silence hepatic targets in non-human primates at the same dose.[5] Consequently, members of our lab sought to identify new compounds that would facilitate delivery at doses less than 1 mg/kg. The original lipoid library demonstrated that fully saturated alkyl tails were more effective than non-alkyl tails and that lipids attached to amines through stable amide bonds were more effective than those connected by ester linkages.[13] To this end, members of our lab developed a second-generation lipidoid library based on stable chemical bonds between fully saturated alkyl tails and amines generated by an epoxide ring-opening reaction[20] (Fig. 7.3). Alkyl tail length was varied between 9 and 18 carbons, and amines were selected if they were effective in previous studies or if they possessed an interesting structural element not yet analyzed. Similar to the Michael addition, the epoxide ring-opening reaction does not require purification or protection steps and can be performed in open air. As a result, a structurally diverse library was synthesized quickly for high-throughput analysis.

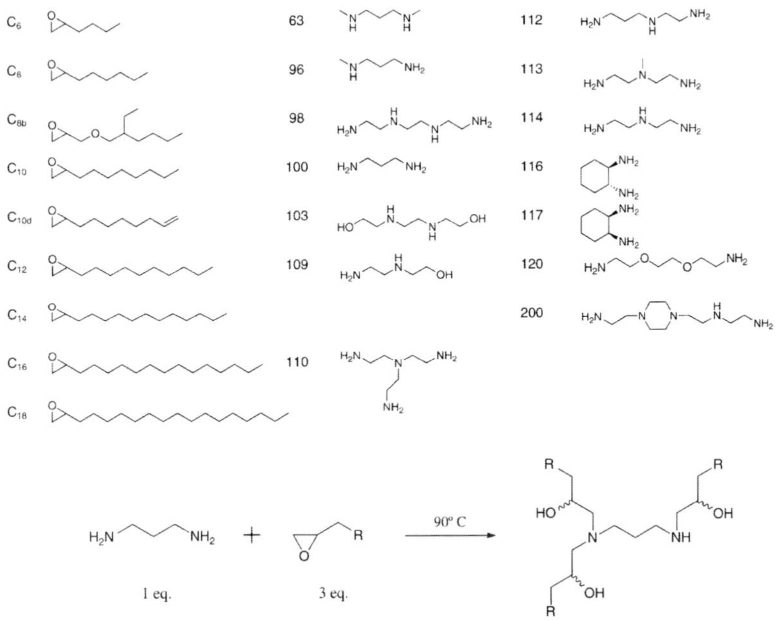

Figure 7.3 Synthesis of epoxide-derived lipidoid library. (Top) Epoxide-terminated alkyl chains and amine-containing monomers were used in synthesis of combinatorial library. (Bottom) Addition of epoxides to amines by efficient ring-opening enables parallel synthesis of library members. Reproduced with permission from Ref. [20].

More than 100 second-generation lipidoids were evaluated *in vitro* using the dual HeLa assay. Several compounds conferred 90% silencing at an siRNA dose of 33 nM. Negligible off-target effects initially measured by constant Renilla expression were later confirmed using an MTS cell-viability assay. Subsequent dose-response studies in HeLa cells demonstrated that three compounds (C14-113, C12-113, and C14-120) silenced the target gene by 70% at 3.3 nM. Based on these studies, 12 promising candidates were tested for their ability to silence F7 *in vivo* at a dose of 3 mg/kg. Three compounds (C12-200, C16-96, and C14-110) silenced F7 completely.

Reduced doses revealed that C12-200 inhibited F7 expression by 50% at a dose of 0.01 mg/kg, increasing efficiency by two orders or magnitude compared to $98N_{12}$-5 and SNALP[5] formulations.

This 100-fold improvement in efficiency was associated with decreased toxicity. Liver enzymes, which are known to be elevated in response to toxicity, remained unaffected, even at a dose of 1 mg/kg. Since drugs are often characterized by the ratio of toxic dose to therapeutic dose (termed the therapeutic window), effective silencing at doses several hundred-fold lower than those generating liver toxicity suggests that this new lead compound, C12-200, might be a clinical candidate. After efficient silencing was achieved in mice, the authors examined hepatocyte silencing in non-human primates. Transthyretin (TTR) was chosen as a target gene, since mutated forms cause familial cardiomyopathy and neuropathy and can be treated only by liver transplant.[21] C12-200 reduced mRNA expression by 90%, 85%, and 75% after doses of 0.3, 0.1, and 0.03 mg/kg, respectively.

C12-200's sizable therapeutic window also enables several genes to be silenced concurrently. The potential to silence many genes at once is one of the most salient advantages siRNA has over traditional small therapeutics. Most pathologies are driven by an assortment of genes acting in concert.[14] For example, viral infections like Hepatitis C evade therapies by rapidly changing gene expression.[22] Similarly, multiple mutations in hepatocellular carcinoma and other cancers are responsible for aggressive growth, metastasis, and drug resistance.[23] Since a single delivery vehicle can transport all of the therapeutic RNAs together, each treated cell would receive an integrated effect. In contrast, small molecule drugs have different pharmacokinetic properties, so some cells might experience the effects of one drug but not another. Accordingly, the authors investigated whether they could silence five hepatic genes related to cholesterol homeostasis. C12-200 was formulated with equal amounts of siRNAs targeting F7, ApoB, proprotein convertase subtilisin/kexin type 9 (PCKS9), sortillin 1 (SORT1), or x-box binding protein (Xpb1). The dose of an individual siRNA was varied between 0.005 and 0.2 mg/kg, leading to a cumulative dose between 0.025 and 1.0 mg/kg. At the highest dose, all five genes were silenced 65–90%. Although this combination was not tested for its potential to treat cholesterol-based pathologies, it did demonstrate that five genes could be knocked down simultaneously (Fig. 7.4).

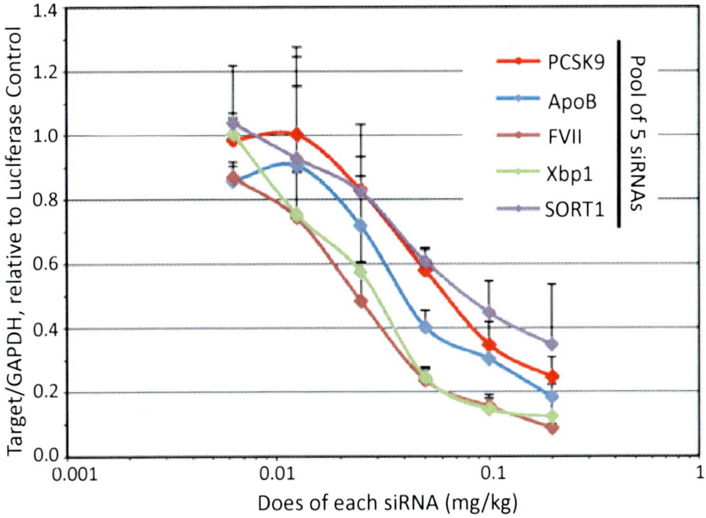

Figure 7.4 Five hepatocellular gene targets were simultaneously silenced by a single injection of pooled siRNAs formulated with C12-200. Mice were administered a single dose and dosage was titrated from 0.2 and .005 mg/kg per siRNA. 72 h post injection, liver tissue was harvested for analysis of gene transcript levels. (s.d., $n = 5$). Reproduced with permission from Ref. [20].

The therapeutic window also enabled us to investigate whether higher siRNA doses resulted in extended silencing. F7 was measured as a function of time after intravenous dosing of mice at 0.1 and 1.0 mg/kg. Serum levels returned to baseline after 20 and 35 days at these respective doses. These results illustrate that siRNA therapies can be tailored to diseases as they evolve over time. For example, as inflammatory responses advance from acute to chronic, the expression levels of the genes responsible for driving disease progression change.[24] A low dose of siRNA targeting innate cell recruitment (important in acute inflammation) could be given in concert with a high dose targeting adaptive cytokine signaling, which is important in both acute and chronic stages.[25] Similarly, cancer cells often upregulate drug efflux pumps and different metabolic genes in response to treatment with chemotherapy.[26] Increased expression of these "contingency" genes allows cells to survive, leading to disease remission. siRNAs targeting traditional oncogenes could be combined with siRNAs targeting such contingency genes, thereby enhancing the efficacy of co-administered small molecule drugs.

7.8 Applications: Using Lipidoids to Treat Disease Models

One of the most noteworthy advantages to RNAi therapeutics is that effective delivery does not depend on RNA sequence. Consequently, the same delivery vehicle can be leveraged to treat many diseases. The best demonstration of this is the effective application of $98N_{12}$-5 to three dissimilar diseases: hypercholesterolemia, malaria, and metastatic prostate cancer. Both hypercholesterolemia and malaria are impacted by hepatocytes, but in disparate ways. Hepatocytes produce and remove cholesterol from the bloodstream, making them critical in cholesterol homeostasis. Malaria, on the other hand, is a multi-stage disease in which small parasites, termed sporozites, mature in the liver before rapidly infecting red blood cells.

$98N_{12}$-5 reduced hypercholesterolemia by silencing PCKS9, a gene whose product binds and degrades low-density lipoprotein receptors (LDLRs) in the liver.[27] LDLRs reduce serum concentrations of this "bad" cholesterol by removing it from the bloodstream. This mechanism, which has been confirmed in mice and humans, is made pathological by somatic or familial mutations that lead to PCKS9 overexpression.[28, 29] Importantly, anti-PCKS9 therapeutics, while successful in many patients, were unable to sufficiently reduce cholesterol levels in others.[27] The combination of a well-defined pathway, a pathology dependent on a gain-of-function mutation, and a patient population unresponsive to traditional therapeutics made PCKS9 an excellent candidate for an RNAi therapy. As a result, we investigated the effect of silencing PCKS9 in normal mice, transgenic mice, rats, and non-human primates.[27] The use of multiple animal models was critical, since rodents transport most of their cholesterol as high-density lipoproteins (HDL), whereas primates transport the majority of the cholesterol as LDL. Such investigations further illustrate the importance of understanding how different species might respond to new therapies.[30]

In addition to testing multiple animal models, it is important to confirm that therapeutic responses are due to RNAi rather than off-target or immunostimulatory effects.[31] Although there is no formal standard procedure to demonstrate this, the case for an on-target RNAi-mediated effect is strengthened by (a) modifying siRNA with non-immunostimulatory 2′-O-methyl groups, (b) testing siRNA for cytokine induction in whole blood or *in vivo*, (c) using a scrambled

RNA control, (d) detecting site-specific cleavage products through 5′-RACE, and (e) confirming that knockdown is dose responsive. While investigating PCKS9 silencing, all of these controls were used. The levels of PCKS9 mRNA, PCKS9 protein, and serum cholesterol were also measured as a function of time. $98N_{12}$-5 reduced PCKS9 and/or serum cholesterol in all four animal models using modified siRNA that did not elicit a cytokine response. In wild-type mice, PCKS9 mRNA was reduced for 28 days after intravenous injections of 7.5 mg/kg, while PCKS9 protein levels were reduced to the limit of detection for three days in transgenic mice after injecting 5 mg/kg siRNA. Serum cholesterol levels remained lowered for 30 days in rats. Target mRNA cleavage was confirmed by 5′-RACE. Delivery of off-target siRNA did not reduce target gene expression, while PCSK9 was silenced in a dose-dependent fashion.

We were curious to examine whether this delivery system could be used to influence not only host pathology but also the infectious disease. Malaria is a multi-stage infection that begins when a mosquito deposits sporozites under the skin of a human host. Sporozites migrate to hepatocytes, attracted by their highly sulfated heparan sulfate proteoglycans.[32] After maturing in the liver, sporozites rupture the hepatocytes, enter the bloodstream, and infect red blood cells. Malaria continues to devastate parts of the developing world, particularly in Africa and Asia. Disturbingly, it was recently reported that mosquitoes may be acquiring resistance to Artemisinin,[33] the most potent anti-malarial drug produced to date. If true, it is projected that the 250 million cases and one million deaths each year caused by malaria will rise dramatically.[33]

To this end, we silenced a hepatic gene that was hypothesized to influence malaria infection, Heme Oxygenase-1 (HO-1).[34] HO-1, which plays a critical role in the metabolism of heme, is normally expressed at low levels. Expression increases in response to several stimuli, including heavy metals, hypoxia, heat shock, and sporozite infection.[34] HO-1 upregulation is known to inhibit cerebral malaria, a dangerous stage of the disease that comes after sporozites relocate from hepatocytes to red blood cells.[35] However, the role of HO-1 in the liver (pre-erythrocytic stage of malaria) remained unclear. To study how HO-1 expression influenced susceptibility to sporozite liver infection, Plasmodium levels were measured in mice expressing HO-1 ($Hmox^{+/+}$) and in mice genetically modified not to express HO-1 ($Hmox^{-/-}$). In contrast to the anti-malarial role it plays in the red blood cell stage of the disease, HO-1 expression was found to increase susceptibility to liver infection. To address this hepatic role of HO-1 therapeutically, siRNA targeting HO-1 was complexed with $98N_{12}$-5 and administered intravenously to $Hmox^{+/+}$ mice at a dose of 5 mg/kg. The resultant 60% silencing of HO-1 achieved through systemic siRNA treatment conferred the same effect as genomic deletion of the gene, namely prevention of blood-stage infection.[34]

This investigation illustrates that genetically modified mice can serve as an important control in siRNA studies. Conversely, in the case that modified mice are not readily available, siRNA-mediated silencing might be used in place of genetic models. In this way, systemic gene silencing can be used to answer interesting biological questions. For example, $98N_{12}$-5 was also used to answer a question critical to the field of RNAi: Whether exogenous siRNAs disrupted endogenous microRNA (miRNA) pathways *in vivo*.[36] miRNAs are small RNAs produced in the nucleus and exported to the cytoplasm, where they regulate genes controlling critical cell processes, including proliferation and survival.[37, 38] Previously, it was reported that genomically integrated short hairpin RNA (shRNA) disrupted miRNA pathways, leading to acute and non-specific toxicity.[39] To investigate whether the introduction of exogenous small RNAs downstream of exportin-5 — which had been saturated in the shRNA experiments, resulting in the observed toxicity — would mitigate this deleterious effect, $98N_{12}$-5 was complexed with siRNA targeting F7 or ApoB and injected intravenously into mice at a dose of either 2 or 5 mg/kg. After confirming target gene silencing, the levels of miRNAs miR-122, miR-16, and let-7 were found to be unaltered relative to saline control, demonstrating that siRNA delivery did not disrupt miRNA production.[36] This was corroborated by the mRNA levels of seven miR-122 targets, which remained unaffected by siRNA delivery.

After demonstrating that intravenously injected lipidoids could efficiently silence hepatic genes, we investigated whether they could be used to silence genes in monocytes[40] and epithelial ovarian cancer cells.[41, 42] Monocytes are precursors to the phagocytotic macrophages of the innate immune system, mediating inflammatory responses to a variety of diseases, including cancer, myocardial infarction, and diabetes.[43] Although paramount to the initial immune response, one subset termed "inflammatory monocytes" often supports disease by promoting chronic inflammation. Such inflammatory monocytes are thus an attractive therapeutic target; however, inactivating them without disrupting other immune functions remained challenging.[44] To inhibit the function of this particular subset, C12-200 was complexed with siRNA targeting chemokine receptor 2 (CCR2), a protein critical in their recruitment.[40] CCR2 mediates inflammatory monocyte behavior but does not have a functional role in noninflammatory monocytes.[40]

Since the spleen, bone marrow, and blood are rich in monocytes, members of our lab quantified real-time biodistribution of C12-200 in multiple organs. After complexation with fluorescently tagged siRNA, particles were tracked using fluorescence molecular tomography and computed tomography (FMT-CT) imaging.[45] siRNA concentration was highest in the monocyte-rich spleen and bone marrow for the first 24 hours following intravenous injection of 1 mg/kg siRNA. Subsequent histological staining and FACS analyses demonstrated that C12-200 transfected inflammatory monocytes more efficiently than other immune cell types. Once delivery to inflammatory monocytes was confirmed, 50% mRNA silencing and reduced protein expression was measured following intravenous injection of 0.5 mg/kg CCR2 siRNA.

By inhibiting inflammatory monocyte recruitment through CCR2 silencing, the authors sought to mitigate inflammatory responses in myocardial infarctions, pancreatic islet transplant rejection, and lymphoma. Since activated macrophages and other inflammatory cells lead both to plaque rupture and increased infarct size once a heart attack occurs, the authors sought to reduce their accumulation in plaques.[46,47] Both macrophage accumulation and infarct size were measured after control or CCR2 siRNA was injected intravenously into mice at a dose of 0.5 mg/kg. Excitingly, both parameters were markedly reduced relative to controls. The authors also investigated whether CCR2 silencing would promote islet transplantation. Although islet transplantation has been successful in treating diabetes, grafts are often rejected by the host immune system.[48] After intravenous injection of lipidoids containing CCR2 siRNA at a dose of 0.5 mg/kg, graft survival was enhanced, leading to improved glycemic performance.

Finally, the authors examined whether CCR2 prompted the recruitment of tumor-associated macrophage (TAMs). TAMs promote tumor growth and metastasis via the release of matrix metalloproteinases (MMPs) and other molecules that degrade extracelluar matrix, releasing growth factors and clearing a path for cells to migrate.[49] Accordingly, the presence of TAMs correlates with poor prognosis in lymphoma.[50] Although metastasis was not used as an endpoint in this study, CCR2 silencing did reduce TAM accumulation, tumor volume, and tumor vascularization after lymphoma cell implantation.

Intravenous administration of siRNA targeting CCR2 at a dose of 0.5 mg/kg resulted in anti-tumor activity, reducing tumor volume and TAM levels by ~50% after five days. However, more than 90% of cancer-related deaths are caused by metastasis, not primary tumors. The small size of the lesions, their propensity to spread to multiple organs, and their biological complexity have traditionally hampered anti-metastatic therapies.[51] Importantly, recent biological studies have identified targets that inhibit disease progression.[52] These genes of interest are often identified via *in vitro* migration assays before cancer cells lacking these genes are surgically implanted and tested for metastatic potential.[53, 54] For example, expression of epiregulin, Cox-2, and MMPs increased vascular permeability, facilitating the migration of breast cancer cell to the lungs.[54] Small molecules are typically tested in these models; however, RNAi may be a better technology for such studies, as silencing persists for days following exposure to the drug. To demonstrate that therapeutic RNAi could reduce metastasis, we used lipidoid-complexed siRNA to silence Claudin-3 (CLDN3) in a metastatic ovarian cancer model. CLDN3 protein forms junctions between adjacent epithelial cells, yet its overexpression in ovarian tumors has been correlated to increased motility and MMP production.[55] Relative to control treatments, tumor growth was retarded by intratumoral injection of $98N_{12}$-5-formulated CLDN3 siRNA at a dose of ~1 mg/kg.[42] Similarly, intraperitoneal injections of CLDN3 siRNA at a dose ~7 mg/kg reduced tumor metastasis and increased survival of mice relative to controls.

7.9 Future Directions and Conclusions

In little more than 10 years, RNAi has fundamentally changed the way biologists study the genome and its implications in disease and medicine. Excitingly, the role of RNA continues to broaden. Specific gene silencing via siRNA has been augmented by miRNA-mediated regulation of gene networks and, more recently, regulation by non-coding RNAs.[37, 56] The non-linear and dynamic behavior of gene networks has recently been a focus of biologists working in concert with advanced mathematics. This line of investigation has led to the identification of genes that, despite being upregulated only by a small amount, act as potent nodes in complicated genomic networks.[57] Such advances will likely reveal a plethora of genetic targets in many cell types.

The roles these new targets will play in biology and medicine may be diverse, but they will share an important commonality: Our ability to utilize them will be limited by our ability to deliver them. While other groups have successfully utilized synthetically complex cationic lipids[58] to deliver siRNA *in vivo*, we employed efficient chemistry to synthesize libraries of structurally diverse lipid-like materials. High throughput *in vitro* and *in vivo* assays designed to test for delivery to hepatocytes-enabled iterative studies, through which two potent compounds, $98N_{12}$-5 and C12-200, were discovered.

Through these iterative studies, we identified variables that impact delivery. Combining small amines — which electrostatically condense nucleic acids and facilitate endosomal escape via the proton-sponge effect[12] — with lipid tails that enhance delivery by interacting with the plasma and endosomal membranes improves delivery.[10, 11, 13, 20] Modifying the amount of excipients in the formulation and particle size enhanced delivery significantly.[17] To this end, new libraries consisting of combinations of materials that work through disparate mechanisms could lead to effective vehicles. Prospective studies could also evaluate combinations of endogenous compounds, combining molecules that are naturally endocytosed[59] with cell-penetrating peptides.[60] Unsuccessful individual structures from these libraries might also be combined to produce a potent vehicle.[19] Once an effective structure is identified, the systematic study of material: siRNA ratios and the effect of excipients can improve the therapeutic window. This is critical since a large therapeutic window permits concurrent silencing of multiple genes simultaneously.[20] Since pathology-inducing genes can vary between patients, RNAi treatments might one day be tailored to patient-specific mutations. This flexibility could be extended to diseases that evolve over time, since high doses of siRNA silence for longer periods (temporal targeting).

Since delivery does not depend on RNA sequence, a successful delivery vehicle can be used to treat sundry diseases driven by upregulation of gene expression in a given cell type. Such versatility was demonstrated with $98N_{12}$-5, which delivered siRNAs to address pathologies as diverse as cancer,[42] malaria,[34] and hypercholesterolemia.[27] While potent and durable gene silencing has been established upon delivery of siRNA to hepatocytes, realizing the full potential of RNAi will require the synthesis of materials and development of appropriate *in vivo* screening assays to facilitate the identification of new compounds that avoid uptake by hepatocytes and confer silencing in other cell types at low doses. Serum-based

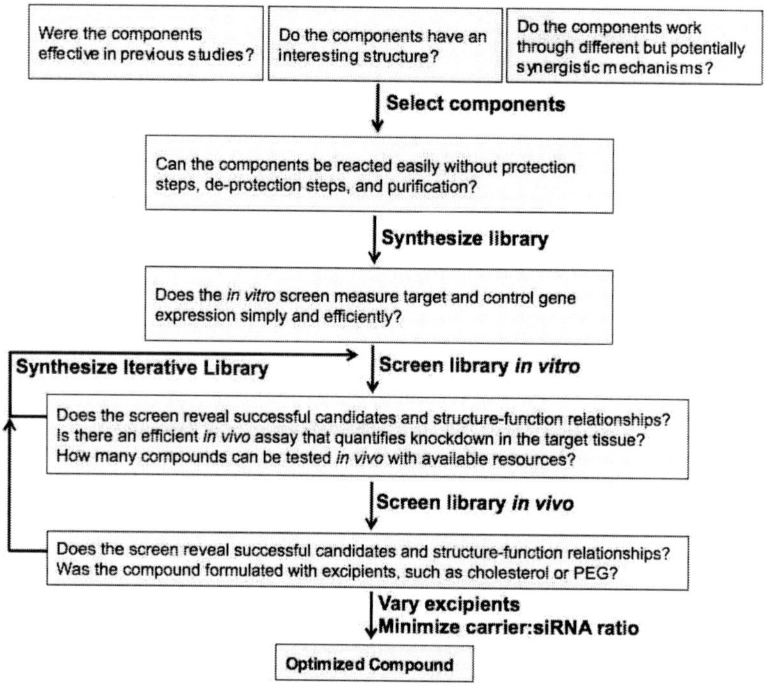

Figure 7.5 Iterative library design. Starting components should be selected rationally, based on previous studies or interesting chemical structure. After ensuring the components can be conjugated using efficient chemistry, efficient *in vitro* screens should be used to identify successful candidates. The top-performing materials can be analyzed to determine structure–function relationships. These results should inform the synthesis of iterative libraries and the selection of compounds screened *in vivo*. Once a potent *in vivo* compound is identified, any excipients used in its formulation should be systematically varied, and the minimum carrier:siRNA mass ratio can be determined.

tests analogous to the F7 assay might identify tissue-specific delivery vehicles, particularly in endocrine organs. However, an ideal assay would measure delivery to every tissue simultaneously and inexpensively. A recently reported mouse model meets these criteria; mice were genetically modified to express luciferase wherever knockdown of a transcriptional repressor occurs.[61] Though spatial resolution using whole-body bioluminescence is somewhat limited, this assay is amenable to high-throughput studies, as the efficacy of many delivery vehicles can be investigated in parallel without sacrificing the mice and harvesting their organs,[62] which is another effective approach to assess pharmacodynamics, albeit at lower throughput. Such improved *in vivo* assays, and others like it, will help uncover structure–function relationships governing the important objective of non-liver delivery. The iterative, combinatorial approach described herein (Fig. 7.5) represents one rigorous approach to achieve this aim.

References

1. Fire, A., *et al.* Potent and specific genetic interference by double-stranded RNA in Caenorhabditis elegans. *Nature*, **391**, 806–811 (1998).
2. Whitehead, K.A., Langer, R., and Anderson, D.G. Knocking down barriers: advances in siRNA delivery. *Nat Rev Drug Discov*, **8**, 129–138 (2009).
3. Peer, D., *et al.* Nanocarriers as an emerging platform for cancer therapy. *Nat Nanotechnol*, **2**, 751–760 (2007).
4. Soutschek, J., *et al.* Therapeutic silencing of an endogenous gene by systemic administration of modified siRNAs. *Nature*, **432**, 173–178 (2004).
5. Zimmermann, T.S., *et al.* RNAi-mediated gene silencing in non-human primates. *Nature*, **441**, 111–114 (2006).
6. Morrissey, D.V., *et al.* Potent and persistent *in vivo* anti-HBV activity of chemically modified siRNAs. *Nat Biotechnol*, **23**, 1002–1007 (2005).
7. Anderson, D.G., Lynn, D.M., and Langer, R. Semi-automated synthesis and screening of a large library of degradable cationic polymers for gene delivery. *Angew Chem Int Ed*, **42**, 3153–3158 (2003).
8. Anderson, D., Akinc, A., Hossain, N., and Langer, R. Structure/property studies of polymeric gene delivery using a library of poly(β-amino esters). *Mol Ther*, **11**, 426–434 (2005).

9. Zugates, G.T., *et al.* Rapid optimization of gene delivery by parallel end-modification of poly(β-amino ester)s. *Mol Ther*, **15**, 1306–1312 (2007).
10. Hafez, I.M., Maurer, N., and Cullis, P.R. On the mechanism whereby cationic lipids promote intracellular delivery of polynucleic acids. *Gene Ther*, **8**, 1188–1196 (2001).
11. Akinc, A., Thomas, M., Klibanov, A.M., and Langer, R. Exploring polyethylenimine-mediated DNA transfection and the proton sponge hypothesis. *J Gene Med*, **7**, 657–663 (2005).
12. Sonawane, N.D., Szoka, F.C., Jr., and Verkman, A.S. Chloride accumulation and swelling in endosomes enhances DNA transfer by polyamine-DNA polyplexes. *J Biol Chem*, **278**, 44826–44831 (2003).
13. Akinc, A., *et al.* A combinatorial library of lipid-like materials for delivery of RNAi therapeutics. *Nat Biotechnol*, **26**, 561–569 (2008).
14. Kumar, V., Abbas, A., Fausto, N., and Aster, J. *Robbins and Cotran Pathologic Basis of Disease*, Eighth Edition, Saunders, Philadelphia, PA (2009).
15. Klibanov, A.L., Maruyama, K., Torchilin, V.P., and Huang, L. Amphipathic polyethyleneglycols effectively prolong the circulation time of liposomes. *FEBS Lett*, **268**, 235–237 (1990).
16. Senior, J., and Gregoriadis, G. Stability of small unilamellar liposomes in serum and clearance from the circulation: the effect of the phospholipid and cholesterol components. *Life Sci*, **30**, 2123–2136 (1982).
17. Akinc, A., *et al.* Development of lipidoid-siRNA formulations for systemic delivery to the liver. *Mol Ther*, **17**, 872–879 (2009).
18. Mahon, K.P., *et al.* Combinatorial approach to determine functional group effects on lipidoid-mediated siRNA delivery. *Bioconjug Chem*, **21**, 1448–1454 (2010).
19. Whitehead, K.A., *et al.* Synergistic silencing: combinations of lipid-like materials for efficacious siRNA delivery. *Mol Ther*, **19**, 1688–1694 (2011).
20. Love, K.T., *et al.* Lipid-like materials for low-dose, *in vivo* gene silencing. *Proc Natl Acad Sci USA*, **107**, 1864–1869 (2010).
21. Sekijima, Y., Kelly, J.W., and Ikeda, S. Pathogenesis of and therapeutic strategies to ameliorate the transthyretin amyloidoses. *Curr Pharm Des*, **14**, 3219–3230 (2008).
22. Kronke, J., *et al.* Alternative approaches for efficient inhibition of hepatitis C virus RNA replication by small interfering RNAs. *J Virol*, **78**, 3436–3446 (2004).

23. El-Serag, H.B. Hepatocellular carcinoma. *N Engl J Med*, **365**, 1118–1127 (2011).

24. Pober, J.S., and Sessa, W.C. Evolving functions of endothelial cells in inflammation. *Nat Rev Immunol*, **7**, 803–815 (2007).

25. Kopf, M., Bachmann, M., and Marsland, B. Averting inflammation by targeting the cytokine environment. *Nat Rev Drug Discov*, **9**, 703–718 (2010).

26. Fletcher, J.I., Haber, M., Henderson, M.J., and Norris, M.D. ABC transporters in cancer: more than just drug efflux pumps. *Nat Rev Cancer*, **10**, 147–156 (2010).

27. Frank-Kamenetsky, M., *et al.* Therapeutic RNAi targeting PCSK9 acutely lowers plasma cholesterol in rodents and LDL cholesterol in nonhuman primates. *Proc Natl Acad Sci USA*, **105**, 11915–11920 (2008).

28. Abifadel, M., *et al.* Mutations in PCSK9 cause autosomal dominant hypercholesterolemia. *Nat Genet*, **34**, 154–156 (2003).

29. Horton, J.D., Cohen, J.C., and Hobbs, H.H. Molecular biology of PCSK9: its role in LDL metabolism. *Trends Biochem Sci*, **32**, 71–77 (2007).

30. Shultz, L.D., Ishikawa, F., and Greiner, D.L. Humanized mice in translational biomedical research. *Nat Rev Immunol*, **7**, 118–130 (2007).

31. Whitehead, K., Dahlman, J.E., Langer, R.S., and Anderson, D.G. Silencing or stimulation? siRNA delivery and the immune system. *Annual Rev Chem Biomol Eng* **2**, 77–96 (2011).

32. Coppi, A., *et al.* Heparan sulfate proteoglycans provide a signal to plasmodium sporozoites to stop migrating and productively invade host cells. *Cell Host Microbe*, **2**, 316–327 (2007).

33. Noedl, H., Socheat, D., and Satimai, W. Artemisinin-resistant malaria in Asia. *N Engl J Med*, **361**, 540–541 (2009).

34. Epiphanio, S., *et al.* Heme oxygenase-1 is an anti-inflammatory host factor that promotes murine plasmodium liver infection. *Cell Host Microbe*, **3**, 331–338 (2008).

35. Pamplona, A., *et al.* Heme oxygenase-1 and carbon monoxide suppress the pathogenesis of experimental cerebral malaria. *Nat Med*, **13**, 703–710 (2007).

36. John, M., *et al.* Effective RNAi-mediated gene silencing without interruption of the endogenous microRNA pathway. *Nature*, **449**, 745–747 (2007).

37. Carthew, R.W., and Sontheimer, E.J. Origins and mechanisms of miRNAs and siRNAs. *Cell*, **136**, 642–655 (2010).

38. Rana, T.M. Illuminating the silence: understanding the structure and function of small RNAs. *Nat Rev Mol Cell Biol*, **8**, 23–36 (2007).
39. Grimm, D., *et al.* Fatality in mice due to oversaturation of cellular microRNA/short hairpin RNA pathways. *Nature*, **441**, 537–541 (2006).
40. Leuschner, F., *et al.* Therapeutic siRNA silencing in inflammatory monocytes in mice. *Nat Biotechnol*, **9**, 1005–1010 (2011).
41. Goldberg, M.S., *et al.* Nanoparticle-mediated delivery of siRNA targeting Parp1 extends survival of mice bearing tumors derived from Brca1-deficient ovarian cancer cells. *Proc Natl Acad Sci USA*, **108**, 745–750 (2011).
42. Huang, Y.-H., *et al.* Claudin-3 gene silencing with siRNA suppresses ovarian tumor growth and metastasis. *Proc Natl Acad Sci*, **106**, 3426–3430 (2009).
43. Gordon, S., and Taylor, P.R. Monocyte and macrophage heterogeneity. *Nat Rev Immunol*, **5**, 953–964 (2005).
44. Charo, I.F., and Ransohoff, R.M. The many roles of chemokines and chemokine receptors in inflammation. *N Engl J Med*, **354**, 610–621 (2006).
45. Nahrendorf, M., *et al.* Hybrid PET-optical imaging using targeted probes. *Proc Natl Acad Sci USA*, **107**, 7910–7915 (2010).
46. Swirski, F.K., *et al.* Identification of splenic reservoir monocytes and their deployment to inflammatory sites. *Science*, **325**, 612–616 (2009).
47. Nahrendorf, M., Pittet, M.J., and Swirski, F.K. Monocytes: protagonists of infarct inflammation and repair after myocardial infarction. *Circulation*, **121**, 2437–2445 (2010).
48. Robertson, R.P. Islet transplantation as a treatment for diabetes – a work in progress. *N Engl J Med*, **350**, 694–705 (2004).
49. de Visser, K.E., Eichten, A., and Coussens, L.M. Paradoxical roles of the immune system during cancer development. *Nat Rev Cancer*, **6**, 24–37 (2006).
50. Steidl, C., *et al.* Tumor-associated macrophages and survival in classic Hodgkin's lymphoma. *N Engl J Med*, **362**, 875–885 (2010).
51. Chiang, A.C., and Massague, J. Molecular basis of metastasis. *N Engl J Med*, **359**, 2814–2823 (2008).
52. Nguyen, D.X., and Massague, J. Genetic determinants of cancer metastasis. *Nat Rev Genet*, **8**, 341–352 (2007).

53. Bos, P.D., *et al.* Genes that mediate breast cancer metastasis to the brain. *Nature*, **459**, 1005–1009 (2009).

54. Minn, A.J., *et al.* Genes that mediate breast cancer metastasis to lung. *Nature*, **436**, 518–524 (2005).

55. Agarwal, R., D'Souza, T., and Morin, P.J. Claudin-3 and claudin-4 expression in ovarian epithelial cells enhances invasion and is associated with increased matrix metalloproteinase-2 activity. *Cancer Res*, **65**, 7378–7385 (2005).

56. Mercer, T.R., Dinger, M.E., and Mattick, J.S. Long non-coding RNAs: insights into functions. *Nat Rev Genet*, **10**, 155–159 (2009).

57. Carro, M.S., *et al.* The transcriptional network for mesenchymal transformation of brain tumours. *Nature*, **463**, 318–325 (2010).

58. Semple, S., *et al.* Rational design of cationic lipids for siRNA delivery. *Nat biotechnol*, **28**, 172–176 (2010).

59. Rajendran, L., Knölker, H.-J., and Simons, K. Subcellular targeting strategies for drug design and delivery. *Nat Rev Drug Discov*, **9**, 29–42 (2010).

60. Morris, M.C., Deshayes, S., Heitz, F., and Divita, G. Cell-penetrating peptides: from molecular mechanisms to therapeutics. *Biol Cell*, **100**, 201–217 (2008).

61. Lin, X., *et al.* A robust *in vivo* positive-readout system for monitoring siRNA delivery to xenograft tumors. *RNA*, **17**, 603–612 (2011).

62. Svensson, R.U., *et al.* Assessing siRNA pharmacodynamics in a luciferase-expressing mouse. *Mol Ther*, **16**, 1995–2001 (2008).

Chapter 8

Manipulation of Leukocytes Using Therapeutic RNAi Delivered by Targeted and Stabilized Nanoparticles

Dan Peer
Laboratory of Nanomedicine, Department of Cell Research and Immunology, and Center for Nanoscience and Nanotechnology, George S Wise Faculty of Life Sciences, Tel Aviv University, Tel Aviv 69978, Israel
peer@post.tau.ac.il

RNA interference (RNAi) has just made it through the pipeline to clinical trials. More studies aiming to translate RNAi toward an ideal therapeutic tool are speeding ahead. However, in order for RNAi to serve as personalized therapeutics and be approved clinically, safe, specific, potent and flexible strategies must be devised for efficient delivery of RNAi payloads to specific cell types. Despite the immense potential, exogenous systemically applied RNAi remains a challenge, particularly the delivery to hematopoietic cells. This chapter describes the current systemic RNAi delivery platforms targeted to leukocytes, with a focus on the integrin-targeted and stabilized nanoparticles (I-tsNPs) strategy, which uses leukocyte integrins for the delivery of siRNAs exclusively to cells of the immune system.

Nanotechnology for the Delivery of Therapeutic Nucleic Acids
Edited by Dan Peer

ISBN 978-981-4411-04-2 (Hardcover), 978-981-4411-05-9 (eBook)
www.panstanford.com

8.1 Introduction

RNAi is a natural cellular mechanism for regulating gene expression that is conserved in a wide range of organisms [1]. This regulation is mediated by dsRNA, which silences the expression of the gene bearing the complementary nucleotide sequences, either by degrading the specific mRNA or by blocking the translation of the mRNA (Fig. 8.1). RNAi is activated by endogenous non-coding RNAs designated microRNAs (miRNAs), which are transcribed initially as pri-miRNA. These long hairpin-structured transcripts are processed by RNase III Drosha into 70 to 100 pre-miRNAs, which are then

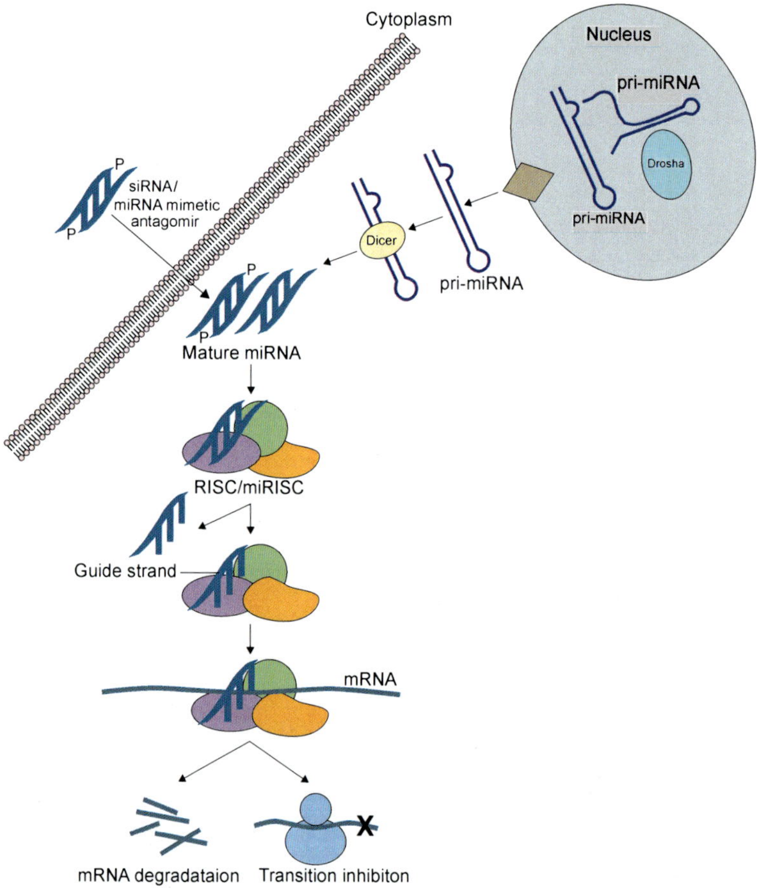

Figure 8.1 The RNAi mechanism for regulating gene expression.

exported to the cytoplasm, where RNase III Dicer generates the mature double-stranded miRNAs of approximately 22 nucleotides. The antisense strand of the mature miRNA is incorporated into the miRNA-containing RNA-induced silencing complex (miRISC), while the other strand is degraded. The miRISC usually hybridizes to partially complementary binding sites on the 3′ untranslated region of target mRNAs. Thus, each individual miRNA can target many different mRNAs, regulating their expression by interfering with translation or by initiating their degradation [2–4].

In the nucleus, the pri-microRNA (miRNA) molecule is processed by the RNase III enzyme Drosha into pre-miRNA. The pre-miRNA is exported into the cytoplasm by Exportin-5, where it is processed by the RNase III enzyme Dicer into the mature miRNA. From this step on, the endogenous pathway is similar to the exogenous pathway; synthetic RNAi molecules (e.g., siRNAs, miRNA mimetics, and antagomirs) are delivered to the cell cytoplasm. The 19 to 23 dsRNA molecule is then incorporated into the RNA-induced silencing complex (RISC), or miRNA-containing RISC (miRISC), the sense strand is released, and the antisense strand mediates the degradation of the target mRNA, or inhibits its translation.

RNAi can be activated by exogenous strategy, for example, by introducing viral vectors expressing shRNA and miRNA duplexes into cells. Alternatively, double-stranded miRNA mimetic molecules, anti-miRNA oligonucleotides (antagomirs), and the frequently used synthetic siRNAs can be taken up directly into the cytoplasm of the cell (Fig. 8.1) [5].

siRNAs are chemically synthesized dsRNAs of 19 to 23 bp with 2-bp nucleotide overhangs at the 3′ ends of each strand. In the cytoplasm of the cell, the siRNA is incorporated into the RISC and undergoes the same process described for endogenous miRNAs. The silencing effect of the RISC complex that incorporates the antisense strand can range from seven days to several weeks in dividing and non-dividing cells, respectively. Moreover, the stable silencing of a compatible mRNA can be achieved by repeated administration of the siRNA. The ability of siRNA to knockdown the expression of any gene of interest makes addressing targets that are not candidates for traditional drug design (e.g., molecules without ligand-binding domains or those sharing structural homology with other molecules) possible. Furthermore, siRNAs that enter a later stage of the endogenous RNAi pathway are less likely to interfere with the gene regulation mediated by the endogenous miRNA machinery; such interference could occur as a result of the saturation of enzymes or transport proteins. In addition, the use of siRNAs for gene silencing eliminates the clinical safety concerns associated with viral vectors. All of these factors make siRNA molecules a promising platform for RNAi-based therapy [4].

Despite the promising therapeutic potential, the biggest challenge that hinders the translation of siRNAs into clinical application is delivery [4, 6, 7]. For example, naked siRNAs cross the cell membrane with low efficiency because of their large molecular weight (~13 kDa), net negative charge, and hydrophilicity. Furthermore, when injected intravenously, siRNAs are subjected to rapid renal clearance and to degradation by serum RNases, resulting in decreased half-lives *in vivo*. In addition, unless modified, siRNAs are recognized by TLRs, which activate cellular cascades that stimulate the immune system and cause a global suppression effect on gene expression, generating off-target effects and the misinterpretation of outcomes. Therefore, while the local delivery of naked siRNA can be useful for treating certain mucosal or subcutaneous tissue diseases, for systemic administration, suitable delivery systems are required to ensure the siRNA reaches the cytoplasm of target cells and activates the RISC. Fortunately, as the chemical properties of synthetic siRNAs are relatively uniform, optimizing the formulation of a given siRNA is likely to apply to other siRNAs, facilitating the development of an siRNA delivery platform technology. Several carriers have been demonstrated to induce gene silencing via systemic siRNA delivery to the liver, kidney, and solid tumors. However, systemic delivery to target leukocytes remains challenging and less advanced. This chapter describes the current systemic RNAi delivery platforms targeted to leukocytes, with a focus on the I-tsNP strategy [8–10], which uses leukocyte integrins for the delivery of siRNAs exclusively to cells of the immune system [11].

8.2 Strategies for RNAi Delivery into Leukocytes

The efficient systemic delivery of siRNAs for gene silencing in leukocytes has great potential for the treatment of a wide spectrum of diseases involving leukocytes, including inflammation, viral infection, and cancer. However, leukocytes and hematopoietic cancer cells are among the most challenging targets for siRNA delivery, as these cells are resistant to conventional transfection reagents and are dispersed in the body, making it challenging to achieve the successful localization of siRNA to the target cells or to deliver siRNAs passively via systemic administration. Therefore, active targeted delivery systems are being developed, in which specific antibodies, ligands, and ligand mimetics are used to mediate the targeting and internalization of the siRNA payloads into cells (Fig. 8.2). Some of these strategies have been demonstrated to deliver siRNAs into leukocytes *in vitro*, including PEGylated nanoscale liposomes with ligands that bind to receptors expressed on leukocytes covalently

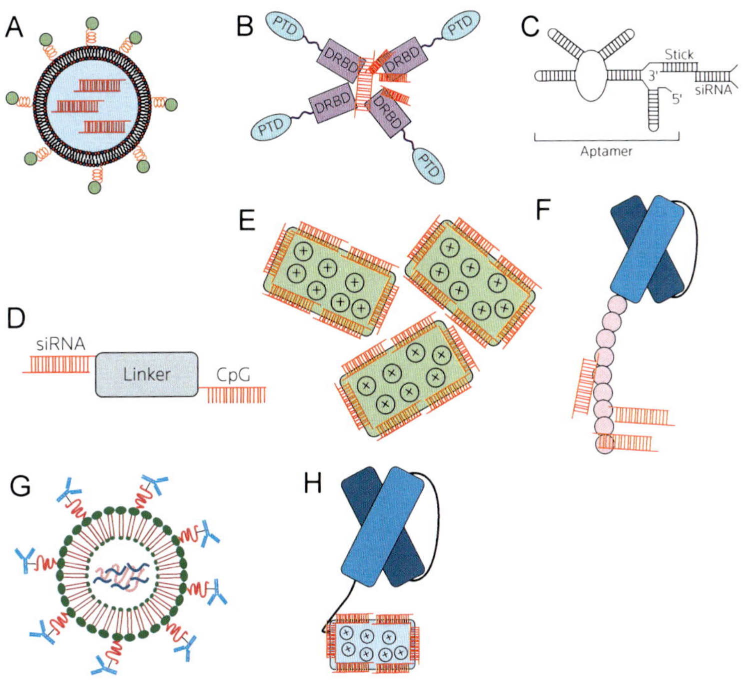

Figure 8.2 Targeted delivery systems for leukocytes. (A) siRNA encapsulated in PEGylated polysome nanoparticles with ligand molecules attached to the PEG; (B) a complex of siRNAs with the PTD-DRBD fusion protein; (C) siRNA synthetically linked to an aptamer; (D) siRNA synthetically conjugated to a CpG oligonucleotide; (E) an atelocollagen-siRNA complex; (F) the 9 arginine (9R) strategy is shown with either an I-tsNP-entrapping siRNA (G), a single-chain variable fragment (scFv)-protamine fusion protein loaded with siRNAs (H), or with a ligand (not shown).

attached to the liposome surface, creating targeting nanoparticles (Fig. 8.2A). In addition, Huang and King [12] described nanoscale liposomes coated with the adhesion molecule P-selectin. Another efficient delivery approach makes use of the peptide transduction domain-dsRNA-binding domain (PTD-DRBD) fusion protein (Fig. 8.2B). In this approach, the DRBD binds to the siRNA, thus masking the negative charge of the molecules, while the PTD mediates the cellular uptake of the complex [13]. Moreover, siRNAs can also be delivered into leukocytes via conjugation to RNA aptamers that bind to specific target molecules, such as glycoprotein 120 [14], without interfering with the binding of the siRNA to Dicer (Fig. 8.2C). In addition to these strategies, other approaches have also been demonstrated to deliver siRNAs into leukocytes efficiently *in vivo* and are discussed in the following sections.

8.3 CpG-Conjugated siRNA

Unmethylated CpG oligonucleotides (Fig. 8.2D) internalize efficiently into dendritic cells, myeloid cells, and B-cells following binding to TLR9. In a recent study by Kortylewski *et al.* [15], CpG oligonucleotides were conjugated to an siRNA targeting the immune suppressor gene *Stat3*. This conjugate was demonstrated to silence *Stat3* in mice, leading to the activation of tumor-associated dendritic cells, macrophages, and B-cells that mediate potent antitumor immune responses, and resulting in tumor cell apoptosis. However, the siRNA in this conjugate was unmodified and negatively charged, reducing its half-life and limiting the therapeutic effect.

8.4 Atelocollagen-Complexed siRNA

Atelocollagen is a type I collagen from calf dermis that does not cause antigenicity or toxicity in animals, as a result of pepsin digestion of the antigenic telopeptides. The positively charged lysine and hydroxylysine residues, which are rich in collagen, are thought to mediate the formation of the atelocollagen–siRNA complex. Ishimoto *et al.* [16] showed that when siRNAs were administrated intravenously with atelocollagen (Fig. 8.2E), the incorporation of the siRNAs into murine macrophages, monocytes, and fibroblasts was facilitated, without causing adverse effects such as the induction of interferon, or liver and renal damage.

8.5 Cationic Nona-D-Arginine Peptide-Complexed siRNA

Kumar *et al.* [17] described a single-chain variable fragment (scFV) form of anti-CD7 that was modified with cysteine at its C-terminus, in order to enable its conjugation to the cationic nona-D-arginine (9R) peptide for the targeted delivery of a negatively charged siRNA payload into mouse T-cells *in vivo* (Fig. 8.2F). This study demonstrated the ability of an siRNA-based treatment to control and suppress HIV replication, without inducing toxicity in target cells. In another study [18], the 9R peptide fused to a dendritic cell-targeting 12-mer peptide delivered siRNA targeting TNFa effectively *in vivo*, in order to suppress the production of TNFa by dendritic cells upon induction in mice. Kim *et al.* [19] reported that TNFa-targeted siRNAs, delivered to macrophages *in vivo* using a 9R peptide fused to the ACh receptor-binding peptide, reduced LPS-induced TNFa levels in the blood and brains of injected mice and led to a significant reduction in neuronal apoptosis.

Despite the promising results obtained from such siRNA delivery approaches, the whole spectrum of hematopoietic cells has not been addressed. In addition, the low payload delivered with these strategies might undermine their clinical efficacy. To overcome these obstacles, a leukocyte cell-specific integrin-based approach using I-tsNPs has been developed by Peer and colleagues (Fig. 8.2G) [8].

8.6 I-tsNP as RNAi Delivery Vehicle for Leukocyte-Associated Diseases

Integrins comprise the largest family of cell adhesion molecules, mediating cell-to-cell and cell-to-extracellular matrix attachment. These molecules are α/β heterodimeric proteins and form at least 24 different α/β heterodimers. Leukocyte integrins mediate the adhesive interaction that is crucial for the migration of cells to sites of inflammation. Antibodies targeting leukocyte integrins have been used for the treatment of inflammatory disorders and function by interfering with the adhesive interactions of leukocytes. Efalizumab (Raptiva) and natalizumab (Tysabri) are examples of humanized, integrin-blocking antibodies that have been approved by the FDA for the treatment of autoimmune diseases. However, in April 2009, efalizumab was withdrawn from markets worldwide because of an increased risk of progressive multifocal leukoencephalopathy.

8.7 Leukocyte Integrins as Targets for siRNA Delivery

The potential use of integrins as receptor targets for RNAi delivery to leukocytes is supported by the following observations: (i) Two of the family members, β_2 and β_7 integrins, are expressed exclusively on leucocytes, enabling the selective targeting of hematopoietic cells; (ii) integrins are constitutively internalized and recycled to enable the migration of leukocytes. Thus, integrin recycling supports the internalization of bound antibodies and peptide, a prerequisite for siRNA-mediated activation of the RNAi pathway; (iii) the most unique feature of integrins is their ability to bind ligands in a dynamically upregulated manner, via a conformational change from a low-affinity conformation in resting cells to a high-affinity conformation upon activation. Various leukocyte-associated diseases are characterized by the high-affinity conformation form of integrins [20, 21]; this observation is expected to increase delivery selectivity by leaving naïve cells untouched. As proof of principle that leukocyte integrins are excellent receptor targets for siRNA delivery to leukocytes, an scFV against the high-affinity form of integrin lymphocyte function-associated antigen 1 (LFA-1) was fused to protamine (Fig. 8.2H). This fusion protein selectively delivered siRNAs into activated lymphocytes, both *in vitro* and *in vivo*.

8.8 The Construction and Characterization of I-tsNP

I-tsNPs are made from neutral unilamellar nanoscale liposomes with a defined size of approximately 80 nm in diameter (Fig. 8.3). The neutral charge of the I-tsNPs eliminates toxic effects, such as the induction of cytokines and liver damage, as caused by cationic lipids and polymers [22]. The unilamellar liposomes are coated with the glycosaminoglycan hyaluronan (HA), which functions as a cryoprotectant [23], stabilizing the particles during lyophilization and rehydration. Moreover, hyaluronan serves as a scaffold for the binding of the antibodies, which endow the targeting capability to the nanoparticles [11]. The encapsulation of siRNAs within the I-tsNPs occurs when the lyophilized particles are rehydrated with water containing protamine-condensed siRNAs. Approximately 80% of the siRNAs are encapsulated, with a payload of approximately 4000 siRNA molecules per particle. I-tsNPs protect the entrapped siRNAs from degradation and lead to efficient silencing, even after pre-incubation with RNase A or 50% serum for 2 h prior to cell transfection. Moreover, when assayed *in vitro*, I-tsNP-encapsulated siRNAs do not induce interferon responses or lymphocyte activation.

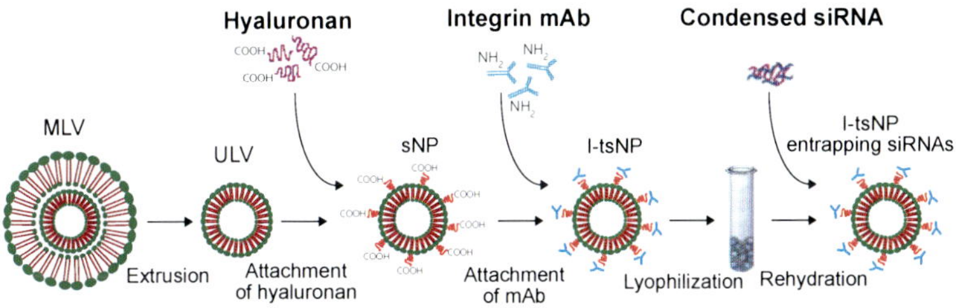

Figure 8.3 The preparation steps of integrin-targeted and stabilized nanoparticles.

Multilamellar vesicles (MLV) are extruded into nanoscale unilamellar vesicles (ULV), which are then surface-modified with high molecular weight (850 kDa) HA to form stabilized nanoparticles (sNPs). The HA-modified liposomes are coupled to the mAb using an amine-coupling method, and then the I-tsNPs are purified and lyophilized. The siRNA entrapment procedure is achieved by rehydrating the lyophilized I-tsNPs with nucleases-free water containing protamine-condensed siRNAs.

8.9 *In vivo* Gene Silencing Using I-tsNP-Entrapping siRNAs

A single intravenous injection of siRNA (2.5 mg/kg) targeted against the DNA repair-associated gene *Ku70* was determined to be sufficient to induce approximately 80% of the target gene expression in β_7 integrin-positive mononuclear leukocytes. In contrast, no silencing effect was observed in mononuclear leukocytes from β_7 integrin-knockout mice that received the same treatment, demonstrating the robust and high targeting specificity of gene silencing *in vivo* using this approach. In addition, the same anti-β_7 I-tsNP platform was used to validate cyclin D1 as an anti-inflammatory target. Cyclin D1 regulates both entry into and progression throughout the cell cycle, and was determined to be upregulated in T-lymphocytes at sites of inflammation, such as in inflamed gut inflammatory bowel disease cells. When β_7 I-tsNP-entrapped cyclin D1 siRNA was intravenously injected into mice with dextran sulphate sodium-induced colitis, cyclin D1 mRNA levels were reduced and led to a notable reduction in intestinal inflammation and tissue destruction. Moreover, in humanized mice, anti-LFA-1-tsNPs carrying siRNAs against the chemokine receptor CCR5 protected mice from HIV infection. Remarkably, the systemic administration of I-tsNP-encapsulated siRNA did not induce interferon responses or lymphocyte activation [24]. The minimal off-target effects and toxicities, in addition to the low doses of siRNA necessary and the high payload capacity, make the delivery of siRNAs via I-tsNPs an economically feasible approach.

8.10 Conclusion

RNAi-based therapy appears to be a promising "magic bullet" for a wide range of diseases, particularly for malignancies. The development of cancer involves alterations in the expression of various genes that are essential for the disease progression. The ability of siRNAs to knockdown the expression of these essential targets might provide a substitute for the highly toxic chemotherapeutic cocktail regimens that are often used to treat cancer. Such RNAi-based therapies are likely to be available in the future, as a result of the significant efforts being made to overcome the challenge of delivering siRNAs to their target tissues *in vivo*. Moreover, all siRNAs possess similar physical and chemical properties; therefore, one effective delivery system can be used as a platform for the delivery of different siRNAs, and by changing the surface antibodies or targeting molecules attached, delivery to different tissues and cells can be achieved. These features are essential for the treatment of leukocyte-associated malignancies, given that in addition to targeting the cancer cells in the peripheral blood system, the hematopoietic cancer cells should also be targeted to prevent the relapse of the disease; this additional targeting requires the use of different siRNAs and targeting molecules. The I-tsNP system that was developed for the delivery of siRNAs to "challenging" hematopoietic cells meets all the requirements from a delivery-vehicle standpoint; as delivery is highly specific and has high payload capability, siRNAs are protected from degradation and toxic effects are not induced at the doses used. Moreover, when administrated systemically, I-tsNP-delivered siRNAs are capable of silencing gene expression with a relatively low amount of siRNAs, making this method economically viable. However, further studies are required to translate this delivery system into the clinical setting.

Acknowledgments

I would like to thank my lab members for helpful discussions. This work was supported in part by grants from the Marie Curie IRG-FP7 of the European Union, Lewis Family Trust, Israel Science Foundation (Award #181/10), the Israel-US Bi-national foundation, the Kenneth Rainin Foundation, the I-CORE Program of the Planning and Budgeting Committee and The Israel Science Foundation (grant No **41/11**), and the FTA: Nanomedicines for Personalized Theranostics of the Israeli National Nanotechnology Initiative to D.P.

References

1. Fire A, Xu S, Montgomery MK, Kostas SA, Driver SE, and Mello CC. Potent and specific genetic interference by double-stranded RNA in Caenorhabditis elegans. *Nature*, 1998; **391**(6669):806–811.
2. Peer D, and Shimaoka M. Systemic siRNA delivery to leukocyte-implicated diseases. *Cell Cycle*, 2009; **8**(6):853–859.
3. de Fougerolles A, Vornlocher HP, Maraganore J, and Lieberman J. Interfering with disease: a progress report on siRNA-based therapeutics. *Nat Rev Drug Discov*, 2007; **6**(6):443–453.
4. Peer D, and Lieberman J. Special delivery: targeted therapy with small RNAs. *Gene Ther*, 2011; **18**(12):1127–1133.

5. Aagaard L, and Rossi JJ. RNAi therapeutics: principles, prospects and challenges. *Adv Drug Deliv Rev*, 2007; **59**(2–3):75–86.
6. Dykxhoorn DM, and Lieberman J. Knocking down disease with siRNAs. *Cell.*, 2006; **126**(2):231–235.
7. Dykxhoorn DM, and Lieberman J. Running interference: prospects and obstacles to using small interfering RNAs as small molecule drugs. *Annu Rev Biomed Eng*, 2006; **8**:377–402.
8. Peer D, Park EJ, Morishita Y, Carman CV, and Shimaoka M. Systemic leukocyte-directed siRNA delivery revealing cyclin D1 as an anti-inflammatory target. *Science*, 2008; **319**(5863):627–630.
9. Elfakess R, and Peer D. Overcoming RNAi transduction in leukocytes using targeted and stabilized nanoparticles. *IDrugs*, 2010; **13**(9): 626–631.
10. Goldsmith M, Mizrahy S, and Peer D. Grand challenges in modulating the immune response with RNAi nanomedicines. *Nanomedicine (Lond)*, 2011; **6**(10):1771–1785.
11. Ben-Arie N, Kedmi R, and Peer D. Integrin-targeted nanoparticles for siRNA delivery. *Methods Mol Biol*, 2012; **757**:497–507.
12. Huang Z, and King MR. An immobilized nanoparticle-based platform for efficient gene knockdown of targeted cells in the circulation. *Gene Ther*, 2009; **16**(10):1271–1282.
13. Eguchi A, Meade BR, Chang YC, Fredrickson CT, Willert K, Puri N, *et al.* Efficient siRNA delivery into primary cells by a peptide transduction domain-dsRNA binding domain fusion protein. *Nat Biotechnol*, 2009; **27**(6):567–571.
14. Zhou J, Swiderski P, Li H, Zhang J, Neff CP, Akkina R, *et al.* Selection, characterization and application of new RNA HIV gp 120 aptamers for facile delivery of dicer substrate siRNAs into HIV infected cells. *Nucleic Acids Res*, 2009; **37**(9):3094–3109.
15. Kortylewski M, Swiderski P, Herrmann A, Wang L, Kowolik C, Kujawski M, *et al. In vivo* delivery of siRNA to immune cells by conjugation to a TLR9 agonist enhances antitumor immune responses. *Nat Biotechnol*, 2009; **27**(10):925–932.
16. Ishimoto T, Takei Y, Yuzawa Y, Hanai K, Nagahara S, Tarumi Y, *et al.* Downregulation of monocyte chemoattractant protein-1 involving short interfering RNA attenuates hapten-induced contact hyper-sensitivity. *Mol Ther*, 2008; **16**(2):387–395.
17. Kumar P, Ban HS, Kim SS, Wu H, Pearson T, Greiner DL, *et al.* T cell-specific siRNA delivery suppresses HIV-1 infection in humanized mice. *Cell*, 2008; **134**(4):577–586.

18. Subramanya S, Kim SS, Abraham S, Yao J, Kumar M, Kumar P, *et al.* Targeted delivery of small interfering RNA to human dendritic cells to suppress dengue virus infection and associated proinflammatory cytokine production. *J Virol*, 2010; **84**(5):2490–2501.

19. Kim SS, Ye C, Kumar P, Chiu I, Subramanya S, Wu H, *et al.* Targeted delivery of siRNA to macrophages for anti-inflammatory treatment. *Mol Ther*, 2010; **18**(5):993–1001.

20. Weinstein S, and Peer D. RNAi nanomedicines: challenges and opportunities within the immune system. *Nanotechnology*, 2010; **21**(23):232001.

21. Peer D. Induction of therapeutic gene silencing in leukocyte-implicated diseases by targeted and stabilized nanoparticles: a mini-review. *J Control Release*, 2010; **148**(1):63–68.

22. Kedmi R, Ben-Arie N, and Peer D. The systemic toxicity of positively charged lipid nanoparticles and the role of toll-like receptor 4 in immune activation. *Biomaterials*, 2010; **31**(26):6867–6875.

23. Peer D, Florentin A, and Margalit R. Hyaluronan is a key component in cryoprotection and formulation of targeted unilamellar liposomes. *Biochim Biophys Acta*, 2003; **1612**(1):76–82.

24. Kim SS, Peer D, Kumar P, Subramanya S, Wu H, Asthana D, *et al.* RNAi-mediated CCR5 silencing by LFA-1-targeted nanoparticles prevents HIV infection in BLT mice. *Mol Ther*, 2010; **18**(2):370–376.

Chapter 9

Lowering the siRNA Delivery Barrier: Alginate Scaffolds and Immune Stimulation

Jana McCaskill, Sherry Wu, Norliana Khairuddin, and Nigel A. J. McMillan*

University of Queensland Diamantina Institute, Princess Alexandra Hospital, Brisbane, QLD 4102, Australia

*n.mcmillan@uq.edu.ac

9.1 Introduction

Since its discovery by Fire and Mello [1] in 1998, RNA interference (RNAi) has made the move from bench to bedside at a breathtaking pace. Indeed, the first clinical trials using RNAi to treat age-related macular degeneration began in 2004, making it one of the most rapidly developed therapeutic agents ever. While seemly ideal for the treatment of cancers, viral and genetic diseases, most RNAi treatments tested in preclinical models have not made it into clinical trials. The key obstacle to harnessing the power of RNAi clinically has been the inability to effectively deliver RNAs to target cells *in vivo*. The overall size, charge, instability, inefficient cell entry, and poor pharmacokinetic profile make delivery challenging, and no universal vector has emerged despite many technologies having been tested. Most of these technologies are nanoparticle based and generally

Nanotechnology for the Delivery of Therapeutic Nucleic Acids
Edited by Dan Peer

ISBN 978-981-4411-04-2 (Hardcover), 978-981-4411-05-9 (eBook)
www.panstanford.com

deliver to the same set of first-pass organs (liver, spleen, and lung) when introduced intravenously. Alternatively, direct delivery into sites such as the eye, skin, and respiratory tract has been used with more success. Once delivery to the organ of interest has been achieved, RNAi molecules must still make contact with target cells and cross the cell membrane to gain access to the cytosol where the RNAi machinery resides. Overcoming these delivery obstacles has been, and will continue to be, challenging with many factors to be considered for the delivery of small interfering RNA (siRNA). Here we outline efforts to improve not only delivery but also RNAi efficacy in the vaginal mucosa as a means to treat genital infections, particularly virally-driven cervical cancer.

9.2 siRNA Delivery Systems: A Brief Overview

Systemic, intravenous, delivery has been the most widely investigated delivery route thus far due to the ease of administration as well as the rapid distribution to various tissue sites. However, it has historically been considered the most complex route of delivery, due to the size, charge, instability, and short half-life of injected siRNAs. This short half-life is caused by renal clearance and endogenous ribonuclease digestion [2]. These obstacles can be reduced by incorporating siRNAs into particles and complexes or by chemically modifying the RNA backbone. The chemical modification of the siRNA backbone through introduction of phosphorothioate and 2′-O-methyl sugar residues shows enhanced resistance toward degradation by exo- and endonucleases in serum and tissue [3]. However, even modified, naked siRNAs have poor cellular uptake due to their small size, net negative charge, renal clearance, and hydrophilicity [4]. Therefore a range of delivery vectors such as liposomes, polymers, and nanoparticles have been developed to facilitate cellular absorption as well as provide a degree of protection and increase systemic circulation time.

9.2.1 siRNA Conjugate Delivery

One method for improved siRNA cellular uptake is through the use of cholesterol conjugates. These siRNA–cholesterol conjugates have been shown to exhibit improved pharmacokinetic profiles and longer *in vivo* half-lives when compared to unconjugated siRNAs [3]. This delivery strategy has been used to deliver siRNAs to treat diabetic nephropathy in a mouse model of type I diabetes [5]. While this study showed good efficacy of siRNA knockdown, it still required a large dose to be delivered to achieve this result.

In addition to cholesterol, siRNAs conjugated to aptamers such as prostate-specific membrane antigen (PSMA, an antigen overexpressed in prostate cancer cells and tumour vascular endothelium) have also been successful in *in vivo* studies targeting prostate tumour cells in a mouse xenograft model [6]. These synthetic aptamers are small, synthetically prepared, highly structured nucleic acid molecules that bind to specific targets. Thus they can facilitate cellular internalization of linked siRNAs by binding to a specific cellular receptor. A key advantage of the aptamer-based system over the cholesterol protein-based system is that the reagents can be produced in a simple *in vitro* transcription reaction that is free of contaminating bacterial or cell products [7].

9.2.2 Peptide-Based Delivery Particles

Peptide-based delivery particles have been used extensively to target siRNAs to specific cell and tissue types. One promising group of cationic peptides is the cell-penetrating peptides (CPPs) that enhance the uptake of a wide range of macromolecules [8, 9]. For example, the rabies virus glycoprotein (RVG), which binds the acetylcholine receptor expressed by neuronal cells, was fused to nine arginines (9R) to allow siRNA binding via ionic interaction [10]. The systemic administration of anti-viral siRNA bound to the chimeric RVG-9R peptide resulted in protection against fatal Japanese encephalitis virus (JEV) infection. Various other CPPs such as HIV-1 TAT and MPG proteins [8, 9], penetratin [11], and polyarginine [12] have been used for the delivery of both protein and nucleic acids into cells.

Another promising peptide delivery modality has been developed by utilising the natural cationic protein, protamine, which binds and condenses negatively charged nucleic acids [13]. Systemic delivery and knockdown has been achieved by wrapping a protamine-condensed siRNA core within cationic lipid membranes that facilitate cellular uptake [14]. Through using these lipid-protamine-siRNA (LPD) nanoparticles, Li and colleagues showed a 70–80% reduction in lung metastatis after two to three doses of LPD particles containing a cocktail of anti-tumour siRNAs targeting VEGF, MDM2, or c-myc [15]. Furthermore, by coupling protamine with antibodies, targeted delivery of siRNA to specific cell types *in vivo* has been shown. Pirollo and colleagues developed nanoimmunoliposome complexes that consisted of a siRNA encapsulated by a cationic liposome that was targeted to primary and metastatic tumours through the surface addition of an anti-TfR single-chain antibody fragment [16].

In the majority of peptide-based delivery systems, non-covalent binding of siRNA and protection from serum nucleases are achieved by simply mixing the siRNA and reagents. This simplicity of peptide-based methods is a great advantage for systemic delivery when compared to other delivery techniques.

9.2.3 Polymer-Based Delivery Vectors

An additional method for systemic siRNA delivery is through polymer-based delivery systems. Polymers such as polyethyleneimine (PEI), chitosan, and poly DL-lactide-co-glycolide (PLGA) serve as efficient transfection reagents due to their ability to bind and condense nucleic acids into stabilized nanoparticles.

PEI is probably the most well-studied polymer delivery system for nucleic acid delivery at local sites as well as systemically. PEI is a synthetic polymer that can form branched or linear forms of different lengths [17] that can be conjugated to siRNA alone or with additional modifications such as PEG or targeting moieties. PEI-conjugated siRNAs have been used to suppress influenza virus infection in mice [18] and knockdown the pain receptor for N-methyl-D-aspartate in rats [19]. Additional conjugations of siRNA PEI nanoparticles with ligands for molecules highly expressed on cancer cells have been achieved for systemic administration. siRNAs targeting survivan, conjugated with PEG and folate were shown to inhibit the growth of transplanted human epidermal carcinoma cells in mice [20]. However, PEI has problems with toxicity upon systemic delivery, although efforts are being made to reduce toxic effects by modifying its structure [21, 22].

An alternative to PEI is chitosan; a natural polysaccharide that has many advantages as a siRNA carrier, including low toxicity, positive charge, and biodegradability [23]. Studies have shown efficient knockdown of tumournecrosis-factor-alpha (TNF-α) expression in macrophages following intraperitoneal administration of chitosan–siRNA nanoparticles for downregulation of local and systemic inflammation [23, 24]. Additionally, chitosan's ability to deliver siRNA across the mucosal epithelium, as seen with intranasal administration of chitosan–siRNA formulations to the lung [25, 26], makes it an attractive delivery platform for a variety of mucosal-based diseases.

Another polymer based delivery system that has received attention recently is PLGA, an FDA-approved, biodegradable polymer that is inherently stable in biological systems. However, this characteristic becomes problematic for endosomal escape and release of siRNA into the cell cytoplasm [27].

9.2.4 Lipid-Based Delivery Particles

Complexing siRNAs into lipoplexes is an alternative method used to assist systemic delivery due to liposomal protection of the siRNA from degradation, reduced renal clearance, and superior uptake abilities [28]. Cationic lipid vectors have been used with success in delivering siRNAs for local application [29], but their use in systemic delivery is limited due to their aggregating properties in serum [30]. Therefore, alternate strategies for systemic lipid delivery have been investigated, including the use of polyethylene glycol (PEG) and neutral lipids, which exhibit superior pharmacokinetic profiles. Utilising PEG and neutral lipids for encapsulation of siRNA molecules into liposomes has been highly successful in systemic knockdown studies [15, 31–33].

A critical factor that influences liposome/siRNA biodistribution is particle size. Particles smaller than 100 nm are required to converse the small liver endothelium and access hepatocytes. Nanoparticles that have a slightly larger diameter of 100–200 nm accumulate in tumours due to the enhanced permeability and retention (EPR) effect. Nanoparticles of this size penetrate into tumours mainly due to a disorganised endothelium and are retained partially due to the lack of lymphatic drainage [34].

In 2005, a scalable method of formulating spontaneous PEGylated lipid particles or stable nucleic acid lipid particles (SNALPs) was developed by Protiva Biotherapeutics [35]. PEGylated siRNA-loaded cationic lipid systems have been widely employed for systemic delivery due to their increased circulatory half-life compared to non-PEGylated liposomes [36]. A modified SNALP method was utilised by Morrissey and colleagues in the earliest demonstration of a systemic anti-viral effect by RNAi using a clinically feasible method. This study showed reduced replication of a hepatitis B virus in mice via IV low pressure injection of chemically modified siRNAs incorporated into SNALPs [37]. SNALPs were further utilised in a study of RNAi-mediated gene silencing in non-human primates [38] and in a proof of concept study by the same research group of post-exposure RNAi therapy of non-human primates against a lethal Ebola virus (EBOV) challenge [39]. Overall, the SNALP system for systemic intravenous delivery is a promising new technique for RNAi therapies.

9.3 HFDM: A Novel Method for Formulating Stable siRNA-Loaded Lipid Particles for *in vivo* Use

One of the issues with liposomes for systemic delivery is the need to complex them with PEG. Addition of PEG is important to ensure liposome accumulation in tissues following intravenous administration. This is due to PEG's ability to create a steric barrier on the surface of liposomes, resulting in decreased interactions with serum proteins. These so-called "stealth liposomes", of which SNALPs are the classic example, are the most widely tested liposomes in clinical trials to date. However, the presence of PEG in the formulation can interfere with the binding efficiency between cationic lipids and anionic siRNA, leading to inefficient entrapment of siRNA in liposomes [40, 41]. More complex procedures have therefore been required to prepare these PEGylated systems compared to those used for conventional lipoplexes. These include the incorporation of repeated freeze-thawing cycles in the formulation procedure [42] or the use of techniques which involve ethanol/detergent dialysis [35, 43], reverse-phase evaporation [44], or post insertion of PEG onto pre-formed lipoplexes [15]. These procedures are labour intensive, time consuming and result in aqueous end products which may not be suitable for long term storage. To address this, we developed a rapid, scalable, and easy method called the hydration-of-freeze-dried-matrix (HFDM) method to produce stealth liposomes for *in vivo* use (Fig. 9.1).

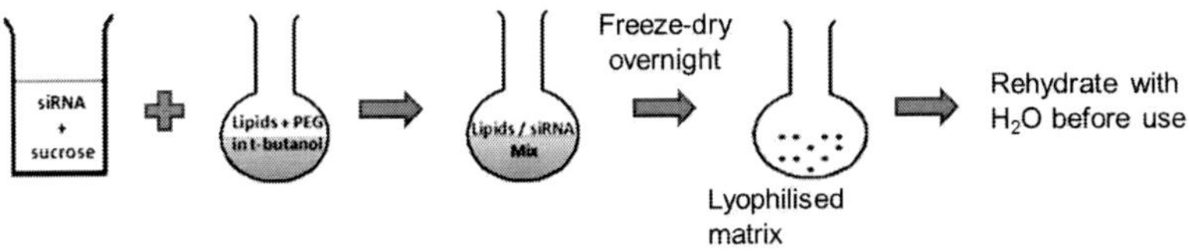

Figure 9.1 Flow chart of the HFDM method for preparing siRNA-loaded lipid particles.

This method holds promise for future systemic RNAi delivery due to its ease of preparation and superior stability of end products [28]. The lipoplexes produced using the HFDM method compare favourably with the commonly used post-insertion technique, producing particles of less than 200 nm in size with high siRNA entrapment efficiency (>90%) and high gene-silencing efficiency (Table 9.1) [28]. Importantly, in contrast to post-insertion preparations, particles made by HFDM retained 100% of their gene-silencing efficiency after storage at room temperature for at least four months. They exhibit a high degree of stability in serum with no breakdown nor aggregation after 24 h [45]. When used *in vivo* via intravenous injection, they rapidly distribute to tissues with the serum concentration dropping to 10% of the initial value within 2 h. The concentration remained steady after 4 h, and the elimination half-life of our particles was 44 h. This is favourable compared to SNALPs, which have an elimination half-life of 6.5 h. When used *in vivo* via intravenous injection, they efficiently accumulate in tumours and are effective at reducing tumour size when targeting the HPV E6/E7 gene in cervical cancer models [45]. Our HFDM particles also

Table 9.1 Summary of size, polydispersity index, zeta potential, and siRNA entrapment efficiency for siRNA-loaded PEGylated lipid particles formulated using either HFDM or post-insertion method

	HFDM method	**Post-insertion method**
Size (nm)	193.2 ± 10.7	178.9 ± 11.8
siRNA entrapment efficiency (%)	94.7 ± 0.61	89.9 ± 2.66
Polydispersity index	0.32 ± 0.03	0.12 ± 0.06
Zeta potential(mV)	45.1 ± 1.17	41.1 ± 1.51

exhibit low toxicity with no elevation of alanine transaminase (ALT) observed, unlike SNALP particles where ALT and aspartate amino transferase (AST) levels were transiently elevated after a single administration of 2.5 mg/kg [38].

Each sample contained 40 μg siRNA in 300 μL isotonic sucrose solution. Lipids used in these formulations include DOTAP, cholesterol, and PEG 2000-C16 Ceramide with a molar ratio of 45:45:10. N/P ratio of 4:1 was used for all formulations. Three batches were made for each formulation method ($n = 3$).

9.4 The Challenge of the Vaginal Tract

Delivery of RNAi mediators to the vaginal tract holds great potential for the treatment of various viral infections responsible for diseases such as acquired immune deficiency syndrome (AIDS), genital herpes, and cervical cancer (reviewed in [46]). Topical vaginal delivery is superior to systemic delivery as it bypasses first pass hepatic and kidney clearance and enables local delivery to the infected target tissue.

However, to be effective as a topical application for the treatment of cancer or viral infections, siRNA molecules must be able to resist rapid degradation by nucleases, inactivation by the low pH vaginal environment, or entrapment in the mucosal layer. Additionally, they must be formulated so that they remain available for cellular uptake by cells targeted by the virus (Fig. 9.2).

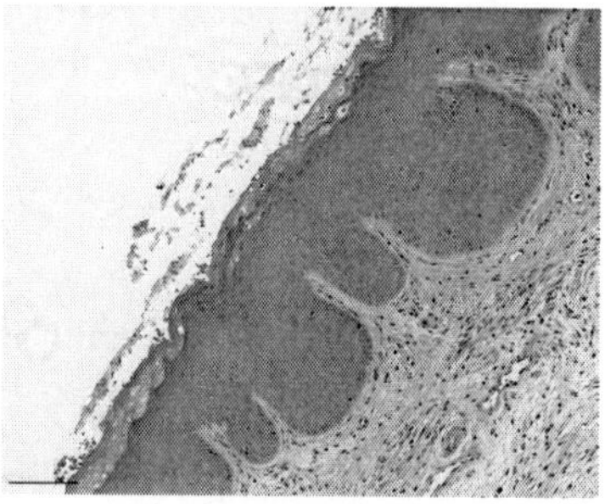

Figure 9.2 Hematoxylin and eosin staining of mouse cervicovaginal section displaying VIN2. siRNAs must be protected from nucleases and the vaginal low pH environment and also be able to penetrate the mucosal layer and outer layer of hyperkeratotic epithelial cells (arrows) to access the dividing tumour cells below. The scale bar represents 100 μM.

The main barrier for efficient vaginal siRNA delivery is the mucosal layer. It is a protective barrier for underlying tissues and removes foreign particles efficiently [47], thus complicating sustained release. An additional factor that may dramatically affect the efficiency of delivery in the vaginal cavity is the physical environment changes that occur throughout the estrous cycle [48].

To date, several groups have reported varying success in delivering siRNA to the vaginal tract. Previous strategies to overcome vaginal delivery barriers have included the simple application of lipoplexes into the vaginal cavity with [49, 50] or without [29] progesterone treatment and mucus removal in the vaginal cavity prior to siRNA administration [51]. Unsurprisingly, investigations into naked siRNA uptake into vaginal tissues after intravaginal administration reported inefficient delivery [29] under normal physiological conditions. This was likely caused by rapid degradation and inefficient mucosal uptake of naked siRNAs in the vaginal cavity. However, the same study showed dramatically improved delivery with the use of the cationic transfection reagent Lipofectamine®. In contrast to this report, we and others have found that conventional lipoplexes were not able to reach the cervicovaginal epithelium and were not retained in the vaginal cavity following administration under normal physiological conditions [52].

Progesterone has been used prior to treatment in multiple studies to arrest mice in the diestrus phase of the estrous cycle. Initial studies that used this method investigated vaginal instillation of siRNAs targeting herpes simplex virus 2 (HSV-2) in a lipid formulation. Palliser and colleagues reported that siRNAs mixed with Oligofectamine® lipid transfection reagent resulted in efficient delivery to the epithelial and lamina propria cells and protected mice from lethal HSV-2 infection [49]. Further investigations into cationic lipidic systems for intravaginal delivery raised concerns regarding their safety *in vivo* [50]. Wu and colleagues reported that inflammation occurred in vaginal tissues following administration of the cationic lipidic transfecting reagent Oligofectamine®. Furthermore, they found that the transfection lipid on its own enhanced transmission of HSV-2. Prevention and protection from HSV-2 was instead obtained by a cholesterol-conjugated siRNA that caused virus gene silencing in the vaginal tract without inflammation or cytokine induction. However, progesterone pre-treatment is typically associated with thinned vaginal epithelia that may facilitate epithelial penetration and is dissimilar to any human estrous phase [53]. Progesterone use has additionally been associated with increased susceptibility and decreased immune responses to vaginal infections [54, 55].

Another strategy of evaluating siRNA silencing in the vaginal tract has used manual removal of the mucus layer [51]. This study also provided an alternative to siRNA lipoplexes with the use of biodegradable polymer-based PLGA nanoparticles. A single dose of siRNA-loaded nanoparticles resulted in efficient and sustained gene silencing from the proximal vaginal lumen to the distal uterine horns.

However, a non-liquid formulation for siRNA vaginal delivery had not been developed. A solid vaginal siRNA delivery system has many benefits over liquid formulations, including the ability to be retained in the vaginal cavity following administration without prolonged anaesthesia, and the capability to be administered at any stage during the estrous cycle. Therefore, our group has developed a more clinically applicable vaginal siRNA delivery platform which could be retained in the vaginal cavity after administration and achieve efficient delivery to the cervicovaginal epithelium under normal physiological conditions.

9.5 Vaginal Delivery of siRNA Using a Novel PEGylated Lipoplex-Entrapped Alginate Scaffold System

To overcome the major limitation of vaginal delivery of siRNAs, we have designed the PEGylated lipoplex-entrapped alginate scaffold (PLAS) delivery system, which is described in detail in our recent paper [52]. As outlined above, the major limitations of past efforts have been poor siRNA or nanoparticle retention time, resulting in reduced efficacy of gene silencing. The mucosal barrier represents the most significant barrier to delivery followed by the hyperkeratotic layer of cells that varies with estrous cycle. The fluid flow and low pH offer additional challenges. Therefore, we have taken the approach of developing a sustained release system based on the entrapment of muco-inert PEG-lipoplexes in a biodegradable scaffold (Fig. 9.3). Thus, we can provide continuous release of siRNA in the vaginal cavity over at least 24 h, and the use of PEG enhances particle stability in the presence of mucin [56]. We chose alginate, a naturally occurring polysaccharide approved by the FDA for various uses, including in food products and wound dressings. It is readily cross-linked by divalent cations such as calcium to form a solid matrix without the use of organic solvents [57, 58]. Scaffolds made from alginate readily decompose in the presence of sodium ions which are naturally occurring within the body and thus serve as a trigger to release entrapped therapeutics slowly over time [59]. Alginate has been previously used in gene delivery either as microspheres [57], poly-ionic complexes [60–62], or hydrogel [58]. Wu *et al.* were the first to report the use of alignate in the form of PLAS for siRNA delivery in the vaginal tract [52].

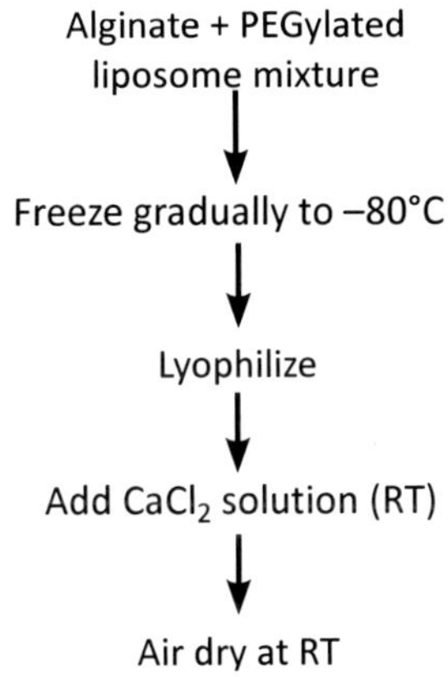

Figure 9.3 Flow chart of the fabrication of PLAS via freeze-dried method.

PLAS was generated by mixing PEG lipoplexes in a 2% solution of alginate in isotonic dextrose and freezing the solution slowly to −80°C before freeze drying (detailed methods are described in [52]). The dried matrix was cross-linked by immersing the matrix in a 5% calcium chloride solution for 3 min before drying overnight. Entrapment efficiencies of lipoplexes were ~50%. Using a mold in the freeze-drying step allows one to achieve any shape PLAS one desires. In the case of the vaginal tract we used 10 mm square sheets that were rolled into cigar-shaped cylinders that made insertion simple and easy. We found that upon exposure to fluids with vaginal pH and sodium characteristics, ~50% of the siRNAs was released within 1 h, the limiting step appearing to be the sodium–calcium exchange. Increasing sodium concentrations resulted in increased siRNA release. The particles retained their particle size and polydispersity.

We found that when inserted intra-vaginally, PLAS offered a six-fold increase in siRNA delivery compared to cationic lipoplexes. More importantly, we were able to achieve 85% knockdown of lamin A/C in vaginal epithelial cells using PLAS [52]. In a cervical cancer model system (C57BL/6 mice expressing HPV16 E6 and E7 under the keratin 14 promoter), we showed that delivery of siRNA against E6/E7 resulted in a significant loss of cells undergoing division, as measured by BrdU incorporation (Fig. 9.4). While the mechanism of alginate-only decrease in proliferation is currently unclear, a significant decrease was observed in siE6/7 group, suggesting effective siRNA delivery. E6 and E7 have been shown to drive suprabasal epithelial cells into cycle [63]. These cells are normally pushed off the basal layer and initiate differentiation. While E6/E7 siRNA does not alter the level of basal cell growth, as expected, the suprabasal cells in cycle are almost completely absent.

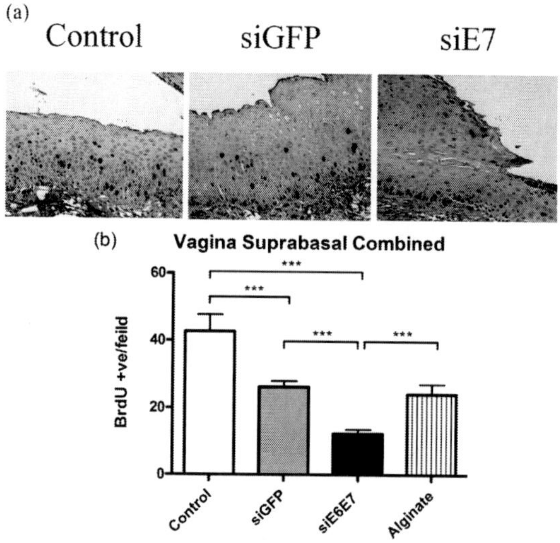

Figure 9.4 Reduced proliferation in the vaginal tract of K14E7 mice upon siRNA treatment. PLAS containing 10 ug of siRNA were inserted into mice pretreated with 5% citric acid for 2 h. Treatment was repeated at days 2 and 4, and mice were sacrificed at day 5.

Overall, PLAS offers a significant improvement in our ability to deliver siRNA in the mucosal setting compared to previous technologies. We envisage PLAS being developed for use in the human vaginal or anal setting for infections such as HSV2 and HPV.

9.6 Thinking Outside the Box: Bi-Functional siRNAs

In addition to improving delivery techniques for RNAi, we have investigated improving the efficacy of RNAi by combining targeted siRNA knockdown with local immune activation. The concept is simple — local immune activation, via triggering of toll-like receptors, in both the viral and tumour setting may offer more potent and beneficial therapeutic effects over siRNA alone. Potent immune activation by dsRNA longer than 30 bp has been clearly documented and leads to cellular shutdown of protein expression. Immune activation occurs as a consequence of dsRNA recognition by the immune system as a common intermediate in many virus life cycles [64]. Originally, 21 bp siRNAs were thought to be too small to be recognised by the sensors and receptors of the innate immune system. However, activation of a robust immune response by siRNAs has now been well established [65]. To date, innate immune activation and subsequent cytokine production by siRNAs have been shown to be predominately mediated by immune cells and caused by recognition via immune sensors of viral infection. The best recognized receptors and sensors of siRNAs are the toll-like receptors 3-7 and 8, retinoic acid inducible gene 1 (RIG-1), melanoma differentiation associated gene 5 (MDA-5), and dsRNA dependent protein kinase (PKR). Each receptor is activated by different motifs and structures in siRNAs. For example, PKR is activated by long (<30 bp) dsRNA and siRNAs [66], while RIG-1/MDA-5 are activated by a 5′-triphosphate [67]. As these molecules form part of the normal anti-viral response, siRNA activation of these receptors and pathways may be advantageous in the treatment of virally induced diseases.

Since it was originally believed that siRNAs were too small to be detected by the immune system, off-target immune activation effects were often poorly controlled in early RNAi studies and many results may have been due more to such activation than RNAi-mediated gene silencing. This was shown to be the case by Tekmira Pharmaceuticals, which demonstrated that in a murine model of influenza infection, the reported anti-viral activity of siRNA was not due to RNAi silencing, but was primarily caused by siRNA immune stimulation [68]. The misinterpretation of anti-viral activity was attributed to the inadvertent low immunogenicity of the control siRNA compared to the anti-influenza siRNA. Indeed, this control siRNA had been used by a number of groups [49, 69–72], and thus this study emphasised the need to carefully consider, monitor, and control for siRNA-mediated immune stimulation.

Immune activation in a subject receiving siRNA treatment remains a major concern when introducing RNAi in a clinical setting. However, it is important to note that non-specific immunostimulatory responses may provide therapeutic benefit in the treatment of certain diseases, particularly in the local delivery setting. Indeed, the potential beneficial aspects of such siRNAs are only now being realized with most studies going to great lengths to avoid any immune stimulation. To date, the use of RNAi to augment immune responses has only been pursued in three ways. Firstly, RNAi has been utilised directly for immunotherapy. For example, anti-IL-10 immunostimulatory siRNAs were shown to both suppress IL-10 expression and activate the production of interferons and cytokines. The activation of these immune mediators simultaneously inhibited the expression of immunosuppressive factors and induced innate immunity to enhance the response of allogeneic T cells [73].

The second application of RNAi immune activation has been for the treatment of viruses. While targeting Influenza A, it was reported that lipoid-delivered unmodified siRNAs had anti-viral activity. However, as siRNAs against unrelated targets also suppressed influenza replication *in vivo*, this activity was not dependent on sequence homology to the influenza genome [74]. Therefore, the majority of anti-viral activity was caused by the immunostimulatory effects of the siRNA and not by its silencing abilities. This suggests that immune activation by RNAi may be beneficial as a therapy for certain viruses. Furthermore, a recent study established that immunostimulatory synthetic 20 bp RNA oligonucleotides have the ability to block suppression by T regulatory cells via activation through RIG-I and MDA-5 [75]. This activation enabled the generation of an efficient immune response against viruses, indicating that immunostimulatory RNAi may initiate multiple pathways that assist in viral knockdown.

Thirdly, immune activating siRNAs have been used as a therapeutic cancer agent. A recent study reported by Poeck and colleagues clearly indicates the potential added benefit of immune activating siRNA in a melanoma cancer model. The activation of RIG-1 by 5′ triphosphate was demonstrated to synergistically promote tumour cell apoptosis when an immunostimulatory siRNA targeting Bcl2 was administered intravenously [76]. Kortylewski and colleagues furthered the initial idea of using immunostimulatory RNAi as an immunotherapy by targeting genes involved in immune inhibition for cancer treatment. This concept was achieved by conjugating a siRNA that targets the immune suppressor gene Stat3 to a CpG oligonucleotide agonist of TLR9. This conjugation enabled three activities in a single molecule: immune cell targeting and TLR activation by CpG and immune checkpoint silencing by the siRNA [77]. The study found that the combination of siRNA silencing and CpG immune activation caused potent anti-tumour immune responses. All of the above studies clearly identify the benefits of siRNA silencing together with the additional benefits of immune activation for therapeutic application.

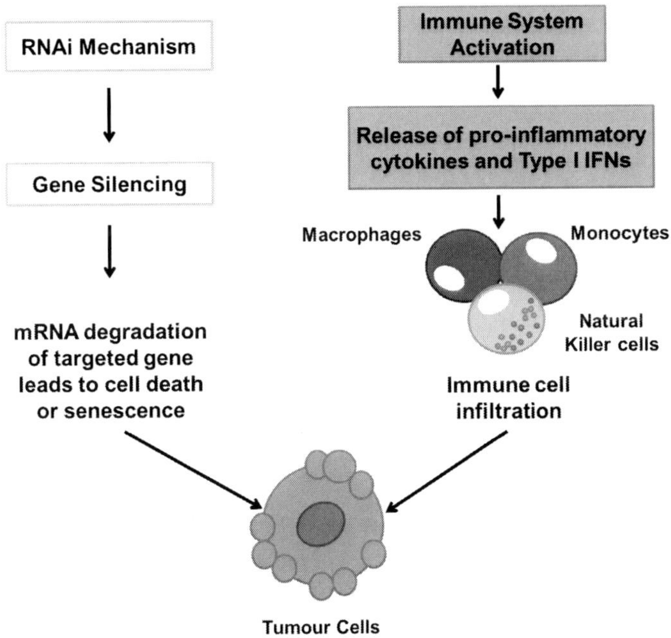

Figure 9.5 Bi-functional RNAi targeting tumour cells. Combined benefit of direct gene silencing from RNAi mechanism and siRNA immune system activation targeting tumour cells.

While immune activation by siRNAs has been readily dismissed in the majority of studies as an unwanted side effect of the silencing agents, the above research indicates that immune stimulation may be beneficial. An emerging concept proposed by our group aims to utilise the ability of siRNAs themselves to activate innate immune responses and design potent "bi-functional siRNAs". Bi-functional siRNAs are so called due to their silencing and TLR-activating abilities, which may be beneficial for viral and cancer therapy (Fig. 9.5). However, the danger upon delivery of bi-functional siRNAs is global or systemic immune activation, perturbing the normal physiology of the host and potentially causing damaging side effects. Therefore, local delivery, and thus activation at the diseased cell or tissue site, would provide the benefits of immune activation without global immune upregulation.

9.7 siRNA-Induced Immunostimulation Promotes Anti-tumoural Activity *in vivo*

To investigate the potential benefits of bi-functional RNAi in the treatment of HPV-driven tumours *in vivo*, we developed a range of unmodified siRNAs targeting the HPV type 16 E6/E7 oncogenes [78]. In this study, we characterized the anti-tumour effects of non-modified siRNAs that can target the specific HPV 16 E6/E7 oncogenes and/or recruit the innate immune system against these HPV-driven tumours. Firstly, these siRNAs were evaluated for gene silencing and immunostimulatory potential *in vivo* upon system delivery of siRNa in our stealth liposomes. We observed that our bi-functional siRNA (IDT4) induced both IL-6 and IFNγ in the serum of mice at 8 h compared to a non-stimulating siRNA (IDT12) or controls (Fig. 9.6).

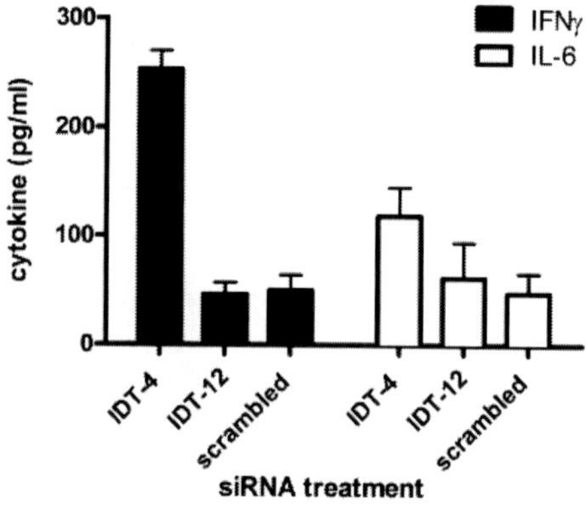

Figure 9.6 C57BL/6 mice were injected with 40 μg siRNA/mouse i.v. and blood collected at 8 h after i.v. Serum from blood was used in detection of pro-inflammatory cytokines in mouse CBA inflammation kit and data read on the FACS Array. Error bar represents ±SEM with $n = 6$.

IDT4 and its modified control, IDT4m7 (which removes immunostimulation) were injected into mice that had previously established HPV-driven tumours. The bi-functional IDT4 was more potent at reducing tumour growth compared to IDT4m7 (Fig. 9.7a). Interestingly, we noted that the use of a highly immunostimulatory, but non-HPV targeting, siRNA was also able to exert anti-tumoural effects at day 10 (Fig. 9.7b), but this effect was not apparent by day 14, suggesting bi-functional siRNA provides superior long-term anti-tumour control (Fig. 9.7b).

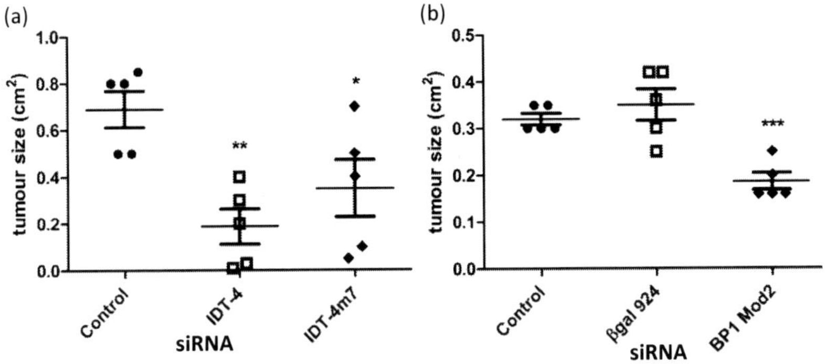

Figure 9.7 Treatments of established tumours with siRNAs show a reduction in tumour growth. TC-1 cells (1×10^6 cells) were subcutaneously injected into C57BL/6 mice on day 0 before mice were treated intravenously with 40 μg siRNAs encapsulated in liposomes on days 5, 8, and 12. (a) Tumour growth reduction by RNAi is independent of TLR7 activation. Treatments carried out with IDT-4 as an immunostimulatory D-siRNA and IDT-4m7 as a non-immunostimulatory D-siRNA, both with RNAi ability for HPV16 E6/E7. Caliper measurements of tumours were carried out on day 14 post initial subcutaneous injections. Error bar represents ±SEM with $n = 6$. Significant differences between control and IDT-4 and control and IDT-4m7 are indicated (*$P < 0.05$, **$P < 0.01$; two-sided t-test). (b) Activation of TLR7 independent of RNAi results in a transient anti-tumour effect. Treatments carried out with 40 μg siRNA mouse per intravenous (i.v.) injection as previously described with BP1 Mod2 (highly immunostimulatory for IFNa) and bgal 924 (non-immunostimulatory for IFNa), both not targeting HPV16 E6/E7. Caliper measurements of tumours were carried out on day 10 post initial subcutaneous injections. Significant difference between control and BP1 Mod2 is indicated (***$P < 0.001$; two-sided t-test). Error bars represents ±SEM with $n = 5$.

Overall this study showed for the first time that siRNA-mediated innate immune activation unaccompanied by oncogene silencing has significant but transient anti-tumour effects in HPV-driven cancers. Our data suggests that a more potent and sustained anti-tumour effect can be obtained by combining siRNA oncogene silencing with innate immune recruitment. This is further supported by the data showing abolition of the immunostimulation activity reduced anti-tumour activity. With improvements to bi-functionality by incorporation of novel sequence modifications that confer stronger immunostimulatory effects, bi-functional siRNAs can be developed as a single molecule approach with multiple activities. This could be a promising anti-tumour therapeutic strategy, including targeting HPV-driven cancers.

9.8 Conclusion

Clinically, siRNA therapies have proven disappointing. Novel approaches to delivery and efficacy will be required to push RNAi into the clinic. The HFDM and PLAS systems show promise in bridging the bench-to-bedside divide.

References

1. Fire, A., *et al.*, Potent and specific genetic interference by double-stranded RNA in Caenorhabditis elegans. *Nature*, 1998. **391**(6669): pp. 806–811.
2. Gao, S., *et al.*, The effect of chemical modification and nanoparticle formulation on stability and biodistribution of siRNA in mice. *Mol Ther*, 2009. **17**(7): pp. 1225–1233.
3. Soutschek, J., *et al.*, Therapeutic silencing of an endogenous gene by systemic administration of modified siRNAs. *Nature*, 2004. **432** (7014): pp. 173–178.
4. Watts, J.K., G.F. Deleavey, and M.J. Damha, Chemically modified siRNA: tools and applications. *Drug Discov Today*, 2008. **13**(19–20): pp. 842–855.
5. Yuan, H., *et al.*, Effects of cholesterol-tagged small interfering RNAs targeting 12/15-lipoxygenase on parameters of diabetic nephropathy in a mouse model of type 1 diabetes. *Am J Physiol Renal Physiol*, 2008. **295**(2): pp. F605–617.

6. McNamara, J.O., 2nd, *et al.*, Cell type-specific delivery of siRNAs with aptamer-siRNA chimeras. *Nat Biotechnol*, 2006. **24**(8): pp. 1005–1015.

7. Kim, S.S., *et al.*, Strategies for targeted nonviral delivery of siRNAs *in vivo*. *Trends Mol Med*, 2009. **15**(11): pp. 491–500.

8. Schwarze, S.R., *et al.*, *In vivo* protein transduction: delivery of a biologically active protein into the mouse. *Science*, 1999. **285**(5433): pp. 1569–1572.

9. Torchilin, V.P., *et al.*, TAT peptide on the surface of liposomes affords their efficient intracellular delivery even at low temperature and in the presence of metabolic inhibitors. *Proc Natl Acad Sci USA*, 2001. **98**(15): pp. 8786–8791.

10. Kumar, P., *et al.*, Transvascular delivery of small interfering RNA to the central nervous system. *Nature*, 2007. **448**(7149): pp. 39–43.

11. Moschos, S.A., *et al.*, Lung delivery studies using siRNA conjugated to TAT(48–60) and penetratin reveal peptide induced reduction in gene expression and induction of innate immunity. *Bioconjug Chem*, 2007. **18**(5): pp. 1450–1459.

12. Kim, H.K., *et al.*, Enhanced siRNA delivery using cationic liposomes with new polyarginine-conjugated PEG-lipid. *Int J Pharm*, 2010. **392**(1–2): pp. 141–147.

13. Vangasseri, D.P., S.J. Han, and L. Huang, Lipid-protamine-DNA-mediated antigen delivery. *Curr Drug Deliv*, 2005. **2**(4): pp. 401–406.

14. Gao, K., and L. Huang, Nonviral methods for siRNA delivery. *Mol Pharm*, 2009. **6**(3): pp. 651–658.

15. Li, S.D., S. Chono, and L. Huang, Efficient oncogene silencing and metastasis inhibition via systemic delivery of siRNA. *Mol Ther*, 2008. **16**(5): pp. 942–946.

16. Pirollo, K.F., *et al.*, Materializing the potential of small interfering RNA via a tumor-targeting nanodelivery system. *Cancer Res*, 2007. **67**(7): pp. 2938–2943.

17. Shim, M.S., and Y.J. Kwon, Controlled delivery of plasmid DNA and siRNA to intracellular targets using ketalized polyethylenimine. *Biomacromolecules*, 2008. **9**(2): pp. 444–455.

18. Thomas, M., *et al.*, Full deacylation of polyethylenimine dramatically boosts its gene delivery efficiency and specificity to mouse lung. *Proc Natl Acad Sci USA*, 2005. **102**(16): pp. 5679–5684.

19. Tan, P.H., *et al.*, Gene knockdown with intrathecal siRNA of NMDA receptor NR2B subunit reduces formalin-induced nociception in the rat. *Gene Ther*, 2005. **12**(1): pp. 59–66.

20. Guo, S., F. Huang, and P. Guo, Construction of folate-conjugated pRNA of bacteriophage phi29 DNA packaging motor for delivery of chimeric siRNA to nasopharyngeal carcinoma cells. *Gene Ther*, 2006. **13**(10): pp. 814–820.

21. Alshamsan, A., *et al.*, Formulation and delivery of siRNA by oleic acid and stearic acid modified polyethylenimine. *Mol Pharm*, 2009. **6**(1): pp. 121–133.

22. Zintchenko, A., *et al.*, Simple modifications of branched PEI lead to highly efficient siRNA carriers with low toxicity. *Bioconjug Chem*, 2008. **19**(7): pp. 1448–1455.

23. Howard, K.A., *et al.*, Chitosan/siRNA nanoparticle-mediated TNF-alpha knockdown in peritoneal macrophages for anti-inflammatory treatment in a murine arthritis model. *Mol Ther*, 2009. **17**(1): pp. 162–168.

24. Nawroth, I., *et al.*, Intraperitoneal administration of chitosan/dsiRNA nanoparticles targeting TNFalpha prevents radiation-induced fibrosis. *Radiother Oncol*, 2010. **97**(1): pp. 143–148.

25. Howard, K.A., *et al.*, RNA interference *in vitro* and *in vivo* using a novel chitosan/siRNA nanoparticle system. *Mol Ther*, 2006. **14**(4): pp. 476–484.

26. Zhang, W., *et al.*, Inhibition of respiratory syncytial virus infection with intranasal siRNA nanoparticles targeting the viral NS1 gene. *Nat Med*, 2005. **11**(1): pp. 56–62.

27. Tiemann, K., and J.J. Rossi, RNAi-based therapeutics-current status, challenges and prospects. *EMBO Mol Med*, 2009. **1**(3): pp. 142–151.

28. Wu, S.Y., *et al.*, Development of a novel method for formulating stable siRNA-loaded lipid particles for *in vivo* use. *Pharm Res*, 2009. **26**(3): pp. 512–522.

29. Zhang, Y., *et al.*, Engineering mucosal RNA interference *in vivo*. *Mol Ther*, 2006. **14**(3): pp. 336–342.

30. Ogris, M., *et al.*, PEGylated DNA/transferrin-PEI complexes: reduced interaction with blood components, extended circulation in blood and potential for systemic gene delivery. *Gene Ther*, 1999. **6**(4): pp. 595–605.

31. Akinc, A., *et al.*, A combinatorial library of lipid-like materials for delivery of RNAi therapeutics. *Nat Biotechnol*, 2008. **26**(5): pp. 561–569.

32. Judge, A.D., *et al.*, Confirming the RNAi-mediated mechanism of action of siRNA-based cancer therapeutics in mice. *J Clin Invest*, 2009. **119**(3): pp. 661–673.

33. Li, S.D., and L. Huang, Targeted delivery of antisense oligodeoxynucleotide and small interference RNA into lung cancer cells. *Mol Pharm*, 2006. **3**(5): pp. 579–588.

34. Matsumura, Y., and H. Maeda, A new concept for macromolecular therapeutics in cancer chemotherapy: mechanism of tumoritropic accumulation of proteins and the antitumor agent smancs. *Cancer Res*, 1986. **46**(12 Pt 1): pp. 6387–6392.

35. Jeffs, L.B., *et al.*, A scalable, extrusion-free method for efficient liposomal encapsulation of plasmid DNA. *Pharm Res*, 2005. **22**(3): pp. 362–372.

36. Yu, W., *et al.*, A sterically stabilized immunolipoplex for systemic administration of a therapeutic gene. *Gene Ther*, 2004. **11**(19): pp. 1434–1440.

37. Morrissey, D.V., *et al.*, Potent and persistent *in vivo* anti-HBV activity of chemically modified siRNAs. *Nat Biotechnol*, 2005. **23**(8): pp. 1002–1007.

38. Zimmermann, T.S., *et al.*, RNAi-mediated gene silencing in non-human primates. *Nature*, 2006. **441**(7089): pp. 111–114.

39. Geisbert, T.W., *et al.*, Postexposure protection of non-human primates against a lethal Ebola virus challenge with RNA interference: a proof-of-concept study. *Lancet*, 2010. **375**(9729): pp. 1896–1905.

40. Podesta, J.E., and K. Kostarelos, Chapter 17 — Engineering cationic liposome siRNA complexes for *in vitro* and *in vivo* delivery. *Methods Enzymol*, 2009. **464**: pp. 343–354.

41. Buyens, K., *et al.*, Elucidating the encapsulation of short interfering RNA in PEGylated cationic liposomes. *Langmuir*, 2009. **25**(9): pp. 4886–4891.

42. Shi, N., R.J. Boado, and W.M. Pardridge, Receptor-mediated gene targeting to tissues *in vivo* following intravenous administration of pegylated immunoliposomes. *Pharm Res*, 2001. **18**(8): pp. 1091–1095.

43. Maurer, N., *et al.*, Spontaneous entrapment of polynucleotides upon electrostatic interaction with ethanol-destabilized cationic liposomes. *Biophys J*, 2001. **80**(5): pp. 2310–2326.

44. Stuart, D.D., and T.M. Allen, A new liposomal formulation for antisense oligodeoxynucleotides with small size, high incorporation efficiency and good stability. *Biochim Biophys Acta*, 2000. **1463**(2): pp. 219–229.

45. Wu, S.Y., *et al.*, Systemic delivery of E6/7 siRNA using novel lipidic particles and its application with cisplatin in cervical cancer mouse models. *Gene Ther*, 2011. **18**(1): pp. 14–22.

46. Cristofaro, P., and B. Ramratnam, RNAi tackles a sexually transmitted disease. *Nat Biotechnol*, 2006. **24**(1): pp. 48–49.

47. Cone, R.A., Barrier properties of mucus. *Adv Drug Deliv Rev*, 2009. **61**(2): pp. 75–85.

48. Kanazawa, T., *et al.*, Effects of menstrual cycle on gene transfection through mouse vagina for DNA vaccine. *Int J Pharm*, 2008. **360**(1–2): pp. 164–170.

49. Palliser, D., *et al.*, An siRNA-based microbicide protects mice from lethal herpes simplex virus 2 infection. *Nature*, 2006. **439**(7072): pp. 89–94.

50. Wu, Y., *et al.*, Durable protection from herpes simplex virus-2 transmission following intravaginal application of siRNAs targeting both a viral and host gene. *Cell Host Microbe*, 2009. **5**(1): pp. 84–94.

51. Woodrow, K.A., *et al.*, Intravaginal gene silencing using biodegradable polymer nanoparticles densely loaded with small-interfering RNA. *Nat Mater*, 2009. **8**(6): pp. 526–533.

52. Wu, S.Y., *et al.*, Vaginal delivery of siRNA using a novel PEGylated lipoplex-entrapped alginate scaffold system. *J Control Release*, 2011. **155**(3): pp. 418–426.

53. Khanna, K.V., *et al.*, Vaginal transmission of cell-associated HIV-1 in the mouse is blocked by a topical, membrane-modifying agent. *J Clin Invest*, 2002. **109**(2): pp. 205–211.

54. Gillgrass, A.E., *et al.*, Prolonged exposure to progesterone prevents induction of protective mucosal responses following intravaginal immunization with attenuated herpes simplex virus type 2. *J Virol*, 2003. **77**(18): pp. 9845–9851.

55. Kaushic, C., *et al.*, Progesterone increases susceptibility and decreases immune responses to genital herpes infection. *J Virol*, 2003. **77**(8): pp. 4558–4565.

56. de la Fuente, M., *et al.*, Nanoparticles as protein and gene carriers to mucosal surfaces. *Nanomedicine (Lond)*, 2008. **3**(6): pp. 845–857.

57. Mittal, S.K., *et al.*, Immunization with DNA, adenovirus or both in biodegradable alginate microspheres: effect of route of inoculation on immune response. *Vaccine*, 2000. **19**(2–3): pp. 253–263.

58. Krebs, M.D., O. Jeon, and E. Alsberg, Localized and sustained delivery of silencing RNA from macroscopic biopolymer hydrogels. *J Am Chem Soc*, 2009. **131**(26): pp. 9204–9206.

59. LeRoux, M.A., F. Guilak, and L.A. Setton, Compressive and shear properties of alginate gel: effects of sodium ions and alginate concentration. *J Biomed Mater Res*, 1999. **47**(1): pp. 46–53.

60. Lim, S.H., I.C. Liao, and K.W. Leong, Nonviral gene delivery from nonwoven fibrous scaffolds fabricated by interfacial complexation of polyelectrolytes. *Mol Ther*, 2006. **13**(6): pp. 1163–1172.

61. Hong, H.J., *et al.*, Accelerated wound healing by smad3 antisense oligonucleotides-impregnated chitosan/alginate polyelectrolyte complex. *Biomaterials*, 2008. **29**(36): pp. 4831–4837.

62. Patnaik, S., *et al.*, PEI-alginate nanocomposites as efficient *in vitro* gene transfection agents. *J Control Release*, 2006. **114**(3): pp. 398–409.

63. Brake, T., *et al.*, Comparative analysis of cervical cancer in women and in a human papillomavirus-transgenic mouse model: identification of minichromosome maintenance protein 7 as an informative biomarker for human cervical cancer. *Cancer Res*, 2003. **63**(23): pp. 8173–8180.

64. Takeuchi, O., and S. Akira, Innate immunity to virus infection. *Immunol Rev*, 2009. **227**(1): pp. 75–86.

65. Marques, J.T., and B.R. Williams, Activation of the mammalian immune system by siRNAs. *Nat Biotechnol*, 2005. **23**(11): pp. 1399–1405.

66. Clemens, M.J., PKR—a protein kinase regulated by double-stranded RNA. *Int J Biochem Cell Biol*, 1997. **29**(7): pp. 945–949.

67. Hornung, V., *et al.*, 5′-Triphosphate RNA is the ligand for RIG-I. *Science*, 2006. **314**(5801): pp. 994–997.

68. Robbins, M., *et al.*, Misinterpreting the therapeutic effects of small interfering RNA caused by immune stimulation. *Human Gene Ther*, 2008. **19**(10): pp. 991–999.

69. Flynn, M.A., *et al.*, Efficient delivery of small interfering RNA for inhibition of IL-12p40 expression *in vivo*. *J Inflamm (Lond)*, 2004. **1**(1): p. 4.

70. Ge, Q., *et al.*, Inhibition of influenza virus production in virus-infected mice by RNA interference. *Proc Natl Acad Sci USA*, 2004. **101**(23): pp. 8676–8681.

71. Song, E., *et al.*, Antibody mediated *in vivo* delivery of small interfering RNAs via cell-surface receptors. *Nat Biotechnol*, 2005. **23**(6): pp. 709–717.

72. Tompkins, S.M., *et al.*, Protection against lethal influenza virus challenge by RNA interference *in vivo*. *Proc Natl Acad Sci USA*, 2004. **101**(23): pp. 8682–8686.

73. Furset, G., and M. Sioud, Design of bifunctional siRNAs: combining immunostimulation and gene-silencing in one single siRNA molecule. *Biochem Biophys Res Commun*, 2007. **352**(3): pp. 642–649.

74. Nguyen, D.N., *et al.*, Drug delivery-mediated control of RNA immunostimulation. *Mol Ther*, 2009. **17**(9): pp. 1555–1562.

75. Anz, D., *et al.*, Immunostimulatory RNA blocks suppression by regulatory T cells. *J Immunol*, 2010. **184**(2): pp. 939–946.

76. Poeck, H., *et al.*, 5′-Triphosphate-siRNA: turning gene silencing and Rig-I activation against melanoma. *Nat Med*, 2008. **14**(11): pp. 1256–1263.

77. Kortylewski, M., *et al.*, *In vivo* delivery of siRNA to immune cells by conjugation to a TLR9 agonist enhances antitumor immune responses. *Nat Biotechnol*, 2009. **27**(10): pp. 925–932.

78. Khairuddin, N., *et al.*, siRNA-induced immunostimulation through TLR7 promotes antitumoral activity against HPV-driven tumors *in vivo*. *Immunol Cell Biol*, 2012. **90**(2): pp. 187–196.

Index